AF361641

Radiation Tolerant Electronics, Volume II

Radiation Tolerant Electronics, Volume II

Editor

Paul Leroux

MDPI • Basel • Beijing • Wuhan • Barcelona • Belgrade • Manchester • Tokyo • Cluj • Tianjin

Editor
Paul Leroux
Electrical Engineering (ESAT)
KU Leuven
Geel
Belgium

Editorial Office
MDPI
St. Alban-Anlage 66
4052 Basel, Switzerland

This is a reprint of articles from the Special Issue published online in the open access journal *Electronics* (ISSN 2079-9292) (available at: www.mdpi.com/journal/electronics/special_issues/radiation_tolerant_v2).

For citation purposes, cite each article independently as indicated on the article page online and as indicated below:

LastName, A.A.; LastName, B.B.; LastName, C.C. Article Title. *Journal Name* **Year**, *Volume Number*, Page Range.

ISBN 978-3-0365-6445-6 (Hbk)
ISBN 978-3-0365-6444-9 (PDF)

Contents

About the Editor . . . . **vii**

Preface to "Radiation Tolerant Electronics, Volume II" . . . . **ix**

Paul Leroux
Radiation-Tolerant Electronics
Reprinted from: *Electronics* **2022**, *11*, 3017, doi:10.3390/electronics11193017 . . . . **1**

Yaqing Chi, Chang Cai, Ze He, Zhenyu Wu, Yahao Fang and Jianjun Chen et al.
SEU Tolerance Efficiency of Multiple Layout-Hardened 28 nm DICE D Flip-Flops
Reprinted from: *Electronics* **2022**, *11*, 972, doi:10.3390/electronics11070972 . . . . **5**

Shuang Jiang, Shibin Liu, Hongchao Zheng, Liang Wang and Tongde Li
Novel Radiation-Hardened High-Speed DFF Design Based on Redundant Filter and Typical
Application Analysis
Reprinted from: *Electronics* **2022**, *11*, 1302, doi:10.3390/electronics11091302 . . . . **17**

Jung-Jin Park, Young-Min Kang, Geon-Hak Kim, Ik-Joon Chang and Jinsang Kim
A Fully Polarity-Aware Double-Node-Upset-Resilient Latch Design
Reprinted from: *Electronics* **2022**, *11*, 2465, doi:10.3390/electronics11152465 . . . . **25**

Arijit Karmakar, Valentijn De Smedt and Paul Leroux
TID Sensitivity Assessment of Quadrature LC-Tank VCOs Implemented in 65-nm CMOS
Technology
Reprinted from: *Electronics* **2022**, *11*, 1399, doi:10.3390/electronics11091399 . . . . **37**

**Stefan Biereigel, Szymon Kulis, Paulo Moreira, Alexander Kölpin, Paul Leroux and Jeffrey
Prinzie**
Radiation-Tolerant All-Digital PLL/CDR with Varactorless LC DCO in 65 nm CMOS
Reprinted from: *Electronics* **2021**, *10*, 2741, doi:10.3390/electronics10222741 . . . . **51**

Karel Appels and Jeffrey Prinzie
Novel Full TMR Placement Techniques for High-Speed Radiation Tolerant Digital Integrated
Circuits
Reprinted from: *Electronics* **2020**, *9*, 1936, doi:10.3390/electronics9111936 . . . . **67**

**Solomon Mamo Banteywalu, Getachew Bekele, Baseem Khan, Valentijn De Smedt and Paul
Leroux**
A High-Reliability Redundancy Scheme for Design of Radiation-Tolerant Half-Duty Limited
DC-DC Converters
Reprinted from: *Electronics* **2021**, *10*, 1146, doi:10.3390/electronics10101146 . . . . **77**

Maria Muñoz-Quijada, Luis Sanz and Hipolito Guzman-Miranda
A Virtual Device for Simulation-Based Fault Injection
Reprinted from: *Electronics* **2020**, *9*, 1989, doi:10.3390/electronics9121989 . . . . **97**

**Tomasz Rajkowski, Frédéric Saigné, Kimmo Niskanen, Jérôme Boch, Tadec Maraine and
Pierre Kohler et al.**
Comparison of the Total Ionizing Dose Sensitivity of a System in Package Point of Load
Converter Using Both Component- and System-Level Test Approaches
Reprinted from: *Electronics* **2021**, *10*, 1235, doi:10.3390/electronics10111235 . . . . **111**

Tomasz Rajkowski, Frédéric Saigné and Pierre-Xiao Wang
Radiation Qualification by Means of the System-Level Testing: Opportunities and Limitations
Reprinted from: *Electronics* **2022**, *11*, 378, doi:10.3390/electronics11030378 . . . . **123**

Jinjin Shao, Ruiqiang Song, Yaqing Chi, Bin Liang and Zhenyu Wu
TAISAM: A Transistor Array-Based Test Method for Characterizing Heavy Ion-Induced
Sensitive Areas in Semiconductor Materials
Reprinted from: *Electronics* **2022**, *11*, 2043, doi:10.3390/electronics11132043 . . . . **139**

Zongru Li, Christopher Jarrett Elash, Chen Jin, Li Chen, Jiesi Xing and Zhiwu Yang et al.
Comparison of Total Ionizing Dose Effects in 22-nm and 28-nm FD SOI Technologies
Reprinted from: *Electronics* **2022**, *11*, 1757, doi:10.3390/electronics11111757 . . . . **149**

Fei Chu, Hongzhuan Chen, Chunqing Yu, Lihua You, Liang Wang and Yun Liu
Quantitative Research on Generalized Linear Modeling of SEU and Test Programs Based on
Small Sample Data
Reprinted from: *Electronics* **2022**, *11*, 2242, doi:10.3390/electronics11142242 . . . . **161**

About the Editor

Paul Leroux

Paul Leroux is Full Professor in the Department of Electrical Engineering (ESAT) of KU Leuven (University of Leuven), Belgium. He received M.Sc. and Ph.D. degrees in electronic engineering from KU Leuven in 1999 and 2004, respectively. From November 2011 to July 2016, he headed the Electrical Engineering Technology Cluster. Since August 2016, he has been the Campus Chair of KU Leuven, Geel Campus. His current research activities within the ESAT-ADVISE research group focus on radiation-hardened analog, mixed-signal, and RF IC design for communication and instrumentation in space, high-energy physics, and nuclear energy applications. His group is part of the CERN CMS collaboration, where Prof. Leroux is the KU Leuven team leader. Prof. Leroux has (co-)authored over 250 papers in international journals and conference proceedings. In 2010, he received the SCK-CEN Prof. Roger Van Geen Award from the FWO/FNRS for his highly innovative work on IC design for harsh radiation environments.

Preface to "Radiation Tolerant Electronics, Volume II"

Research on radiation tolerant electronics has increased rapidly over the last few years, resulting in many interesting approaches to model radiation effects and design radiation hardened integrated circuits and embedded systems. This research is strongly driven by the growing need for radiation hardened electronics for space applications, high-energy physics experiments such as those on the large hadron collider at CERN, and many terrestrial nuclear applications, including nuclear energy and safety management. With the progressive scaling of integrated circuit technologies and the growing complexity of electronic systems, their ionizing radiation susceptibility has raised many exciting challenges which are expected to drive research in the coming decade.

After the success of the first Special Issue on Radiation Tolerant Electronics, the current Special Issue features thirteen articles highlighting recent breakthroughs in radiation tolerant integrated circuit design, fault tolerance in FPGAs, radiation effects in semiconductor materials and advanced IC technologies and modelling of radiation effects.

Paul Leroux
Editor

electronics

Editorial

Radiation-Tolerant Electronics

Paul Leroux 

Department of Electrical Engineering (ESAT), KU Leuven, Kleinhoefstraat 4, 2440 Geel, Belgium;
paul.leroux@kuleuven.be

Citation: Leroux, P.
Radiation-Tolerant Electronics.
Electronics **2022**, *11*, 3017. https://
doi.org/10.3390/electronics11193017

Received: 20 September 2022
Accepted: 21 September 2022
Published: 22 September 2022

Publisher's Note: MDPI stays neutral
with regard to jurisdictional claims in
published maps and institutional affiliations.

When thinking of radiation-tolerant electronics, many readers will think of space. Indeed, with the rise of New Space and Space 2.0 and the corresponding vast growth in space satellites and space vehicles, the need for radiation-tolerant electronics has increased beyond the typical NASA and ESA space missions. New custom radiation-tolerant electronics are needed and more validation and qualification strategies are required for off-the-shelf components. Even beyond space, the need for radiation-tolerant electronics has increased tremendously, for example, applications in aerospace; in high-energy physics such as the Large Hadron Collider experiments at CERN; in upcoming nuclear fusion reactors such as ITER (International Thermonuclear Experimental Reactor) and other fusion reactor technologies; to improve safety in current nuclear energy facilities; or for nuclear waste processing, storage, or transport. Additionally, even beyond these high-energy applications, radiation-tolerant electronics are needed in critical applications such as self-driving cars, where the mean time between failures should be extremely high, and even large data centres or advanced medical devices, where errors—even from a single cosmic particle—can simply not be tolerated.

These abundant applications, together with the evolution of chip technology towards smaller devices that can be upset by progressively less energy, have fuelled research on the fundamentals and modelling radiation effects in electronics, on the design of radiation-tolerant electronics in state-of-the-art technologies, and on new and more efficient ways to evaluate and test the reliability of electronic components in radiation environments.

After the success of the first Special Issue on radiation-tolerant electronics, the current Special Issue features thirteen articles highlighting recent breakthroughs in radiation-tolerant integrated circuit design, fault tolerance in FPGAs, radiation effects in semiconductor materials and advanced IC technologies, and modelling of radiation effects.

Many of the contributions within this Special Issue deal with the design of radiation-tolerant integrated circuits, either at block level or with comprehensive circuits in state-of-the-art IC technologies. Article [1] discusses the SEU (single-event upset) tolerance of three layout-hardened 28 nm DICE (dual interlocked storage cell) D flip-flops implemented in advanced 28 nm planar CMOS technology. In [2], the authors present a cell-level radiation-hardening-by-design (RHBD) method based on commercial processes, showcasing new radiation-hardened D-type flip-flops (DFF) with highly improved SEU tolerance compared to standard DICE flip-flops even with TMR. Article [3] presents a fully polarity-aware double-node-upset (DNU)-resilient latch. The circuit boasts multiple thresholds, an increased number of SEU-insensitive nodes, low power dissipation, and has the strongest radiation-hardening capability among other DNU-resilient latches. In [4], the authors present a comprehensive assessment of TID effects on the performance of a parallel-coupled and super-harmonic-coupled voltage-controlled oscillator (VCO) operating between 2.5 GHz and 2.9 GHz. The circuits are implemented in 65 nm CMOS technology and feature different radiation-hardening techniques. Paper [5] presents the first fully integrated radiation-tolerant all-digital phase-locked loop (ADPLL) and clock and data recovery (CDR) circuit for wireline communication applications. Several radiation-hardening techniques are proposed to achieve state-of-the-art immunity to SEEs up to

62.5 MeV cm^2 mg^{-1} as well as a 1.5 Grad TID tolerance. A final circuit design paper is presented in [6]. This article presents a novel physical implementation methodology for high-speed, triple-modular redundant (TMR), digital-integrated circuits for harsh-radiation-environment applications. An improved distributed approach is presented to constrain redundant branches of TMR digital logic cells using repetitive, interleaved micro-floorplans.

The above paper [6] also fittingly bridges towards two other articles focusing on fault immunity and fault injection on the FPGA (field-programmable gate array) level. In [7], Banteywalu et al. present a high-reliability spatial and time redundancy (TR) hybrid technique, applied to the design of a radiation-tolerant digital controller for a dual-switch forward half-duty limited DC-DC converter. The technique has the potential of double-fault masking with a <2% increase in resource overhead cost compared to TMR, while offering a more than an order of magnitude increase in reliability improvement factor (RIF). Article [8] describes the design and implementation of a virtual device to perform simulation-based fault injection campaigns in existing FPGA (field-programmable gate array)-based hardware devices. Multiple instances of the virtual device can be launched in parallel in order to speed up the fault injection campaigns.

Two articles in this issue describe recent results on the challenging problem of system-level radiation effects' characterization. In [9], Rajkowski et al. compare the system-level evaluation of a point-of-load (PoL) converter under total ionizing dose (TID) with an individual radiation assessment of the different component. It is shown that, due to internal compensation in the system, the complete system can be fully functional at a TID level more than two times higher than the qualification level obtained using a standard-based component-level approach. In continuation of this research, article [10] discusses the opportunities and limitations of radiation qualification by means of system-level testing. To this end, TID and SEE tests are performed and analysed on a system-in-package (SIP) PoL converter. Limitations for the SEE qualification proved substantially stronger than for the TID qualification.

Two articles in this issue deal more with the fundamental effects of radiation in semiconductor materials and advanced IC technologies. In [11], the authors present a transistor-array-based test method for characterizing the heavy-ion-induced sensitive area in semiconductor materials as well as the impact of transistor layout and well contacts for both NMOS and PMOS devices in 65 nm CMOS technology. Article [12] presents a comparison of TID effects in 22 nm and 28 nm FDSOI (fully depleted silicon-on-insulator) technologies. The test structures include ring oscillators designed with inverters, NAND2, and NOR2 gates, as well as SRAM memory cells and flip-flop chains. Overall, the 22 nm FDSOI shows better resilience.

The final article [13] in this issue deals with modelling aspects of radiation effects in complex digital ICs. Different methods are compared for the quantitative evaluation of the SEU cross section under different test programs. A laser test is used to generate training and validation data under these different test programs. The results show that the quantitative evaluation method based on generalized linear models can achieve the highest accuracy.

Funding: This research received no external funding.

Acknowledgments: I would like to thank all of the researchers who submitted articles to this Special Issue for their excellent contributions. I am also grateful to all of the reviewers who helped in the evaluation of the manuscripts and made very valuable suggestions to improve the quality of the contributions. I would like to acknowledge the editorial board of MDPI *Electronics*, who invited me to guest edit this Special Issue. I am also grateful to the *Electronics* editorial office staff who worked thoroughly to maintain the rigorous peer-review schedule and timely publication.

Conflicts of Interest: The author declares no conflict of interest.

References

1. Chi, Y.; Cai, C.; He, Z.; Wu, Z.; Fang, Y.; Chen, J.; Liang, B. SEU Tolerance Efficiency of Multiple Layout-Hardened 28 nm DICE D Flip-Flops. *Electronics* **2022**, *11*, 972. [CrossRef]
2. Jiang, S.; Liu, S.; Zheng, H.; Wang, L.; Li, T. Novel Radiation-Hardened High-Speed DFF Design Based on Redundant Filter and Typical Application Analysis. *Electronics* **2022**, *11*, 1302. [CrossRef]
3. Park, J.; Kang, Y.; Kim, G.; Chang, I.; Kim, J. A Fully Polarity-Aware Double-Node-Upset-Resilient Latch Design. *Electronics* **2022**, *11*, 2465. [CrossRef]
4. Karmakar, A.; De Smedt, V.; Leroux, P. TID Sensitivity Assessment of Quadrature LC-Tank VCOs Implemented in 65-nm CMOS Technology. *Electronics* **2022**, *11*, 1399. [CrossRef]
5. Biereigel, S.; Kulis, S.; Moreira, P.; Kölpin, A.; Leroux, P.; Prinzie, J. Radiation-Tolerant All-Digital PLL/CDR with Varactorless LC DCO in 65 nm CMOS. *Electronics* **2021**, *10*, 2741. [CrossRef]
6. Appels, K.; Prinzie, J. Novel Full TMR Placement Techniques for High-Speed Radiation Tolerant Digital Integrated Circuits. *Electronics* **2020**, *9*, 1936. [CrossRef]
7. Banteywalu, S.; Bekele, G.; Khan, B.; De Smedt, V.; Leroux, P. A High-Reliability Redundancy Scheme for Design of Radiation-Tolerant Half-Duty Limited DC-DC Converters. *Electronics* **2021**, *10*, 1146. [CrossRef]
8. Muñoz-Quijada, M.; Sanz, L.; Guzman-Miranda, H. A Virtual Device for Simulation-Based Fault Injection. *Electronics* **2020**, *9*, 1989. [CrossRef]
9. Rajkowski, T.; Saigné, F.; Niskanen, K.; Boch, J.; Maraine, T.; Kohler, P.; Dubus, P.; Touboul, A.; Wang, P. Comparison of the Total Ionizing Dose Sensitivity of a System in Package Point of Load Converter Using Both Component- and System-Level Test Approaches. *Electronics* **2021**, *10*, 1235. [CrossRef]
10. Rajkowski, T.; Saigné, F.; Wang, P. Radiation Qualification by Means of the System-Level Testing: Opportunities and Limitations. *Electronics* **2022**, *11*, 378. [CrossRef]
11. Shao, J.; Song, R.; Chi, Y.; Liang, B.; Wu, Z. TAISAM: A Transistor Array-Based Test Method for Characterizing Heavy Ion-Induced Sensitive Areas in Semiconductor Materials. *Electronics* **2022**, *11*, 2043. [CrossRef]
12. Li, Z.; Elash, C.; Jin, C.; Chen, L.; Xing, J.; Yang, Z.; Shi, S. Comparison of Total Ionizing Dose Effects in 22-nm and 28-nm FD SOI Technologies. *Electronics* **2022**, *11*, 1757. [CrossRef]
13. Chu, F.; Chen, H.; Yu, C.; You, L.; Wang, L.; Liu, Y. Quantitative Research on Generalized Linear Modeling of SEU and Test Programs Based on Small Sample Data. *Electronics* **2022**, *11*, 2242. [CrossRef]

Article

SEU Tolerance Efficiency of Multiple Layout-Hardened 28 nm DICE D Flip-Flops

Yaqing Chi [1], Chang Cai [2,*], Ze He [3], Zhenyu Wu [1], Yahao Fang [1], Jianjun Chen [1] and Bin Liang [1]

[1] College of Computer, National University of Defense Technology, Changsha 410073, China; yqchi@nudt.edu.cn (Y.C.); wuzhenyu@nudt.edu.cn (Z.W.); yhfang@nudt.edu.cn (Y.F.); chenjianjun09@nudt.edu.cn (J.C.); liangbin@nudt.edu.cn (B.L.)

[2] State Key Laboratory of ASIC and System, Fudan University, Shanghai 201203, China

[3] Institute of Modern Physics, Chinese Academy of Sciences, Lanzhou 730000, China; heze@impcas.ac.cn

* Correspondence: caichang@fudan.edu.cn

Abstract: Three layout-hardened Dual Interlocked Storage Cell (DICE) D Flip-Flops (DFFs) were designed and manufactured based on an advanced 28 nm planar technology. The systematic vertical and tilt heavy ion irradiations demonstrated that the DICE structure contributes to radiation tolerance. However, it is hard to achieve immunity from a Single Event Upset (SEU), even when a ~3-μm well isolation is utilized. The SEU mitigation of the hardened DFFs was affected by the data patterns and clock signals due to the imbalance in the number of upset nodes. When the clock signal equalled 0, no error was observed in ^{181}Ta irradiation, indicating that the DICE DFFs are SEU tolerant in vertical irradiation owing to their reasonable isolation of sensitive volumes. The divergences of SEU cross-sections were enlarged by our specially designed joint change of tilt incidences for both the along-cell and cross-cell irradiation of heavy ions. The evaluations of SEU for both the vertical and tilt irradiations assist with eliminating the overestimation of SEU tolerance and guarantee the in-orbit safety of spacecraft in harsh radiation environments.

Keywords: D Flip-Flop; heavy ion; radiation hardened; Single Event Upset

Citation: Chi, Y.; Cai, C.; He, Z.; Wu, Z.; Fang, Y.; Chen, J.; Liang, B. SEU Tolerance Efficiency of Multiple Layout-Hardened 28 nm DICE D Flip-Flops. *Electronics* **2022**, *11*, 972. https://doi.org/10.3390/electronics11070972

Academic Editor: Paul Leroux

Received: 20 February 2022
Accepted: 19 March 2022
Published: 22 March 2022

Publisher's Note: MDPI stays neutral with regard to jurisdictional claims in published maps and institutional affiliations.

1. Introduction

As a key component of digital circuits, there is wide concern about the irradiation tolerance of D Flip-Flops (DFFs) in advanced technologies in the context of the increasingly unparalleled performance requirements of space electronic systems, despite their reduced area and power consumption [1–4]. The standard Dual Interlocked Storage Cell (DICE) has been applied to DFFs in deep-submicron planar Complementary Metal Oxide Semiconductor (CMOS) technologies to achieve low Single Event Upset (SEU) rates [3,4]. However, the critical charges of SEU for DFF cells are not high, especially for the advanced nanoscale technologies [3–6]. Moreover, the heavy ion-induced charge sharing phenomenon among adjacent sensitive nodes increases the probability of upsets, making the basic hardening techniques ineffective [5,6]. Thus, it is essential to characterize the radiation tolerance and evaluate the effectiveness of hardening strategies of nanoscale circuits.

In recent years, some heavy ion irradiation results for the standard and hardened DFFs of different process nodes have been characterized, and the main SEU cross sections in the published literature are shown in Table 1 [3–6]. The heavy ion irradiation results for the 65 nm standard DFF, basic DICE DFF, and a temporal-DICE DFF are presented in Ref. [3]. The temporal-DICE DFF comprises a temporal structure for its master latch and a DICE structure for its slave latch, which is expected to be SEU hardened. However, merely the basic DICE DFF presents an enhanced radiation tolerance, indicating that the basic DICE DFF appears to be the most attractive for achieving a very high radiation hardness with the least circuit overheads in terms of area and power dissipation [3]. The different SEU cross sections of 40 nm and 28 nm DFFs are illustrated in Ref. [4]. It has been found that the standard 40 nm design has a larger SEU cross-section than the standard 28 nm design [4]. At the LET value of 60 MeV·cm^2·mg^{-1}, the upset cross sections for the 28 nm designs are statistically identical, whereas there is still a noticeable improvement for the 40 nm capacitive hardened DFF, indicating that for the advanced technologies, using capacitance to reduce SEU cross-sections for high LET particles is unattractive [4]. The SEU cross-sections for a broad scope of parameters including the clock frequency and angle of incidence are characterized for the hardened and unhardened DFFs in 32 nm Silicon on Isolator (SOI) technology [5]. The 32 nm DICE DFF is improved in SEU tolerance, while the influence of LET values and frequency is significant. Additionally, the LET values used in Ref. [5] are not high and, therefore, cannot fully characterize the failure rates of hardened DFFs. Thus, the tilt incidence of high-LET heavy ions should be utilized to further investigate the mechanisms of SEU sensitivities, especially for the hardened DFFs. Besides, the vertical heavy ion irradiations are utilized to study the standard and hardened DFFs that employ compact (1.05 μm) or separate (2.25 μm) DICE structures in 22 nm SOI technology in Ref. [6]. Additionally, the enhanced SEU tolerances are verified for the DICE DFFs with either compact or separate structures [6]. Thus, the spacing of sensitive nodes is also an essential parameter that affects the SEU tolerance of DICE DFFs.

Table 1. Heavy ion irradiation results of DFFs with different process nodes of planar technologies in published Refs. [3–6]. The LET (MeV·cm^2·mg^{-1}) values and the corresponding cross-sections (σ: cm^2·bit^{-1}) are shown below. (Cap = capacitance; Stan. = Standard DFF without radiation hardening; T-DICE = Temporal-DICE).

	65 nm			40 nm		32 nm SOI	
Type	**Stan.**	**DICE**	**T-DICE**	**Stan.**	**Cap. (×2.5)**	**Stan.**	**DICE**
Max. LET	~48	~48	~48	~60	~60	~40	~40
σ	~3.3 × 10^{-6}	~6.7 × 10^{-8}	~1.6 × 10^{-6}	~1.1 × 10^{-8}	~1.5 × 10^{-8}	~1.5 × 10^{-10}	~6.5 × 10^{-11}

	28 nm			22 nm SOI		
Type	Stan.	Cap. (×1.5)	Cap. (×3)	Stan.	DICE (1.05 μm)	DICE (2.25 μm)
Max. LET	~60	~60	~60	~85	~85	~85
σ	~4.5 × 10^{-9}	~4.5 × 10^{-9}	~4.5 × 10^{-9}	~7.5 × 10^{-10}	~5.0 × 10^{-11}	~5.0 × 10^{-11}

Based on the discussions above, it is confirmed that the DICE hardened DFFs with multiple node spaces have not been fully investigated using systematic heavy-ion irradiations. The high-LET heavy-ion irradiations with different tilt angles are not available in the literature, but the high-LET ions are essential for verification of the SEU sensitivities and hardening effects on the circuits [6–11]. In addition, the 28 nm bulk planar devices are not well represented in the layout-hardened circuits and irradiation results, with the result that the SEU mechanisms of 28-nm planar devices are not clear [11–16]. Therefore, the characterization of 28 nm DICE hardened DFFs with different node spaces is essential to reveal the basic features of SEU sensitivities and promote the effective application of radiation hardening design for the 28 nm high-performance digital circuits and systems.

In this paper, three different DICE hardened DFFs were designed and fabricated in a 28 nm planar technology to fully characterize their SEU sensitivities. The test circuits were fabricated using a metal-gate process with a high-k gate dielectric, and isolated with shallow trench isolation (STI) technology. It is expected that the measured radiation results of the designed DFFs will provide sufficient SEU support data to guide the design of in-orbit applications. The rest of the paper is organized as follows: Section 2 details our specially designed DFFs including the layouts with hardening strategies and irradiation parameters; Section 3 presents the results for vertical and tilt irradiation of heavy ions; and in Section 4, the irradiation results are summarized and discussed in detail.

2. Circuits Design and Irradiation Setup

The DFF test chip was designed and fabricated using a commercial 28 nm planar bulk silicon process. The test structure was constructed with three chains of DICE DFFs and each chain contained 2000 cascaded DFFs. The three DFF chains had a shared data input (DI) and clock (CK), but their output ports (DOs) were separate. All the DFFs were designed with the normal DICE structure consisting of the double master latches and double slave latches as shown in Figure 1. The CK buffers employed in each DICE DFF determine the working state of the master latches or the slave latches in the DFF. When CK = 1, the dual interlocked master latches ML1 and ML2 maintain the logic value, while the slave latches SL1 and SL2 are bypassed. When CK = 0, the slave latches SL1 and SL2 maintain the logic value while the master latches ML1 and ML2 are bypassed. All the DFFs in one chain were designed with the same layout structures, but the DFFs in different chains were designed with different layout structures as shown in Figure 2. The three detailed layout structures of DFF0, DFF1 and DFF2 are shown in Figure 2a–c, respectively. The circuit structures and the basic layout of the three DFFs are identical, whereas the different node spacing of the three DFFs was designed to achieve the exact isolation of the two interlocked latches of the DICE structure. The drain regions of the off-state MOS transistors are regarded as sensitive volumes (SV). In addition, the minimum SV spacings of the two interlocked latches in DFF0, DFF1, and DFF2 were ~1.68 μm, ~1.95 μm, and ~3.00 μm, respectively. The effectiveness of DICE hardened circuits and the degree of charge sharing effects for 28 nm planar technology were evaluated directly according to the irradiation results of the different DFFs.

An SEU test system was developed to evaluate the SEU sensitivities of the different DFFs. As shown in Figure 3, the system was composed of a motherboard and a daughterboard. The DFF test chip installed on the daughterboard was controlled and monitored by the FPGA-based motherboard via a digital board-to-board I/O interface. A host computer located in the heavy-ion radiation room was connected to the FPGA via a RS-232 interface in order to control the test and record all the data. Another remote computer located outside the radiation room was connected to the host computer inside the radiation room via an ethernet link to enable the operator conduct the test.

Before the heavy-ion irradiation, the FPGA provided the input value via the DI and a 40 MHz clock signal via the CK to each chain to first initialize the stored logic value of all the DFFs. Then the clock signal CK was set to stable to place the DFFs in a static mode and fix the working latches in a DFF. After that, the heavy ions struck the DFF test chip until the fluence reached 10^7 ions·cm^{-2}. Then the 40 MHz clock signal was inputted to the CK, and the logic value stored in all the DFFs was read by the FPGA, which recognized and counted all the upsets simultaneously for each chain. The upset count of each chain was reported to the host computer, and the host computer analyzed the data in real time, displayed the SEU counts, and recorded all the information.

The heavy-ion tests were conducted with the Single-Event Effect Test Terminal (SEETT) at the Heavy Ion Research Facility in Lanzhou (HIRFL) of the Chinese Academy of Sciences. The flux of ions was controlled at 10^4 ions·cm^{-2}·s^{-1}. The vertical irradiation (0°-tilt) and tilt incidences (30°, 45°, and 60°) were used, and the air-layer and 8.3 μm passivation were accounted for in the calculation of the ions' energy and Linear Energy Transfer (LET) values so that the experimental heavy ions reached the SV with sufficient energy deposition. The heavy-ion irradiation conditions and parameters used in the experiments are listed in Table 2. The DFF test chips were de-capped before exposure to irradiation, and the passivation layers on the top of the chip included aluminum, copper, silicon dioxide, tungsten, and other passivation materials. The Input/Output (I/O) and core voltages of the test chip were set at 1.8 V and 0.8 V, respectively, during the irradiation. The ^{181}Ta ions with a controllable 1865 MeV of energy at the surface of the sample chip were selected to receive the large-tilt incidence during irradiation, and the high-LET ^{181}Ta ions were selected to evaluate the SEU sensitivities of the hardened circuits with isolated SVs. In addition, the X-direction (along-cell irradiation) and Y-direction (cross-cell irradiation) were classified.

Hence, a coefficient for the effective fluence (F_{eff}) of ions in SV was required for the tilt incidence, which is related to the beam fluences (F) counted by the particle detector and the cosine value of tilt angle with a vertical direction (θ).

$$F_{eff} = F \cdot cos\theta \tag{1}$$

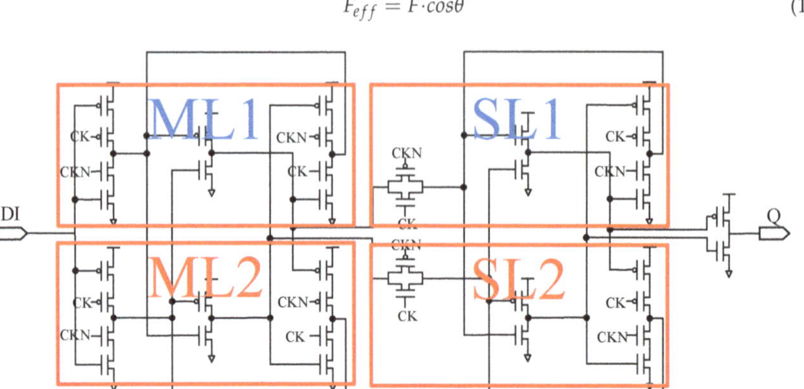

Figure 1. The basic structure of the hardened DFFs.

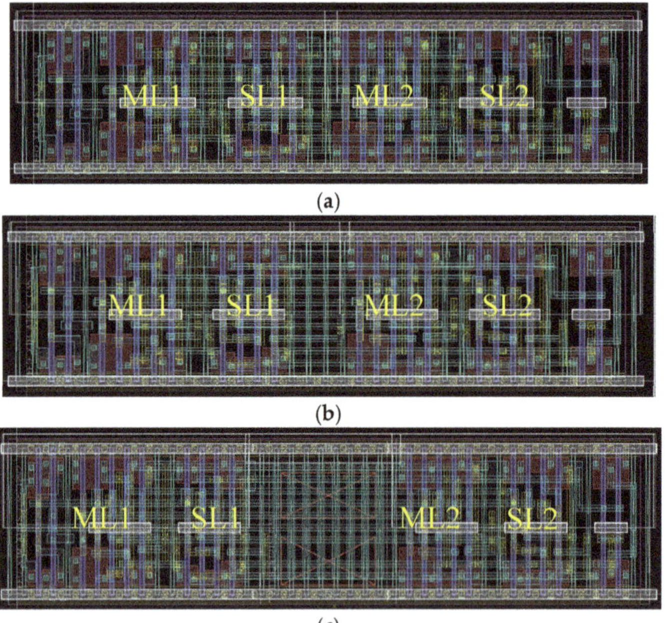

Figure 2. Layouts of (**a**) DFF0 with a minimum sensitive node at ~1.68 μm, (**b**) DFF1 with a minimum at ~1.95 μm (minimum well spacing), and (**c**) DFF2 with a minimum sensitive node at 3.00 μm (well spacing).

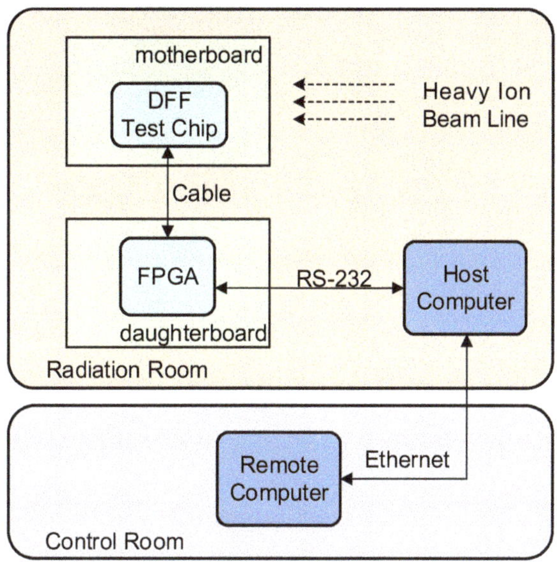

Figure 3. Specially designed SEE testing system.

Table 2. Information about the parameters of ions and irradiation conditions.

Energy in SV (MeV)	LET (MeV·cm²·mg⁻¹)	Range in Silicon (μm)	Tilt (°) (x, y)	Data Pattern	Clock
1695.3	78.3	99.2	(0, 0)	1 & 0	1 & 0
1623.8	79.0	95.3	(0, 45)	1 & 0	1
1521.7	80.1	89.8	(0, 60)	1 & 0	1
1668.8	78.6	97.8	(0, 30)	1 & 0	1
1623.8	79.0	95.3	(45, 45)	1 & 0	1
1623.8	79.0	95.3	(45, 0)	1 & 0	1
1668.8	78.6	97.8	(30, 0)	1 & 0	1
1521.7	80.1	89.8	(60, 0)	1 & 0	1

3. Irradiation Results

The iradiation results are presented in this section. We only recorded the error events induced by a single ion; thus, the SEU cross-sections (σ) are calculated by

$$\sigma = \frac{\sum_j j \cdot N_j}{F \cdot N \cdot cos\theta} \quad j = 1, 2, 3, \cdots \tag{2}$$

where j is the error bits of an SEU event, N_j is the number of SEU events involving j-bit errors, F is the beam fluences, $cos\theta$ is the cosine value of tilt angle with vertical direction, and N is the total bits of DFF. The SEU cross-sections of three DFF chains were extracted by our test system. The one-sigma error bar of SEU cross-sections was calculated for each experimental condition and noted in our following figures. The results of static SEU cross-sections with different data patterns and CK signals are shown in Figure 4. It was found that no Multiple Bit Upsets (MBU) were observed in the test, because each DFF was placed in a unique well region, and all of the DFFs in a chain were separated by over 15 μm, which is effective in preventing the charge sharing effects and MBUs. The downward-pointing arrows in Figure 4 mark the limited value of no upset events. It means that if no upset event was observed during the full irradiation procedure, the maximal SEU cross-sections ($1/(F \cdot N \cdot cos\theta)$) are marked in the figure and labeled with the downward-pointing arrows.

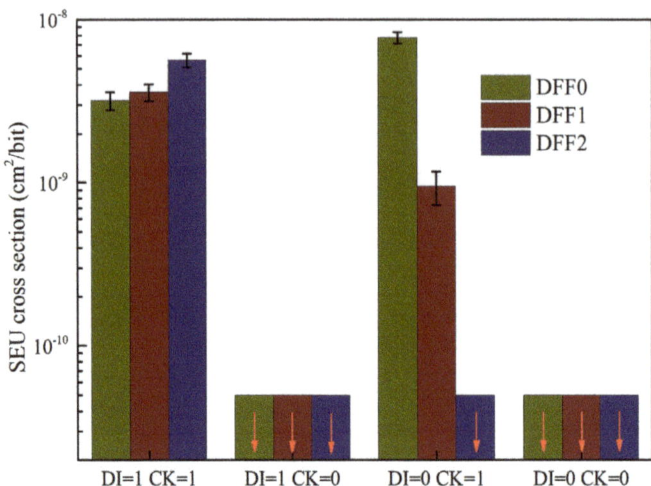

Figure 4. Data patterns and CK dependence of SEU cross-sections.

For CK = 0, the master latches were under the bypass state, while the logic values of slave latches was maintained. No upset was observed in vertical ^{181}Ta irradiations (CK = 0), indicating that the DICE DFFs have SEU tolerance owing to their reasonable isolation of SVs. For CK = 1, the working functions of master latches and slave latches were exchanged. When both the CK and DI were equal to 1, the SEU cross-sections of DFF0, DFF1, and DFF2 were ~3.2 × 10^{-9} cm^2/bit, ~3.6 × 10^{-9} cm^2/bit, and ~5.6 × 10^{-9} cm^2/bit, respectively. When the CK = 1, and the DI = 0, the SEU cross-sections of DFF0, DFF1, and DFF2 were ~7.7 × 10^{-9} cm^2/bit, ~9.5 × 10^{-10} cm^2/bit, and <5.0 × 10^{-11} cm^2/bit, respectively. The SEU cross-sections indicate that the large-area well isolation can improve the SEU tolerance of the DFFs for full 0 data, whereas for the full 1 data, the well isolation seems to slightly increase the SEU sensitivities. The measured SEU cross-sections of the two data patterns are different because the structures of the DFFs are asymmetric. The circuit-level simulations with the double-exponential model were conducted to investigate the imbalance of the SEU susceptibility of DFFs, and the results are shown in Table 3. The number of upset nodes means that the upset data of DFF was observed if the selected nodes of transistors in DFF were injected by the transient pulse with an equivalent LET value at ~80 MeV·cm^2·mg^{-1}. Based on the data in Table 3, it is clear that the number of upset nodes was significantly increased when CK = 1. This is because the DICE structure of the master latch and slave latch in DFF is asymmetric for the consideration of driving behavior. In addition, the asymmetric structure leads the SEU cross-sections of DFFs to have clock and data pattern dependency.

Table 3. Results of the double-exponential pulse injections (LET = ~80 MeV·cm^2·mg^{-1}).

	DFF0	DFF1	DFF2
Number of upset nodes (DI = 0, CK = 0)	2	2	2
Number of upset nodes (DI = 1, CK = 0)	2	2	2
Number of upset nodes (DI = 0, CK = 1)	16	16	16
Number of upset nodes (DI = 1, CK = 1)	13	13	13

The systematic along-cell irradiations with 0–60° changeable tilt angles as well as the cross-cell irradiations with 0–60° changeable tilt angles were all conducted, and the results are shown in Figures 5 and 6. The SEU cross-sections of DFF0 presented in Figure 5a,b indicate that the SEU sensitivities of DICE DFF0 tend to increase with the increase of the tilt angles. However, the tendency for the variation of SEU cross-sections depends on the direction of incidences and data patterns. It is obvious that the full 0 data is sensitive to both the along-cell tilts and cross-cell tilts, while the SEU cross-sections of full 1 data are not improved for the 60° cross-cell irradiation when compared with the 45° cross-cell irradiation, which does not conform to the law of effective LET. In addition, a slight increase of SEU cross-sections with the increase of tilt angles for the full 1 data of the well-isolated DICE DFF1 was observed, as shown in Figure 5c,d, and the improvements of SEU cross-sections for the 60° tilt angle were not obvious compared with the 45° tilt angle. Besides, the hardening effectiveness of the full 0 data for DFF1 decreased in tilt angle irradiations. Moreover, an obvious inhibitory effect of well isolation on the SEU sensitivities was observed in the results presented in Figure 5e,f. The DICE DFF2 showed SEU immunity for the full 0 data until tilt angles over 45°, whereas the 60° cross-cell tilt incidences made the large well isolation less effective. When DI = 1, the SEU cross-sections of DFF0-2 at 60° tilt were two to three times larger than at vertical irradiation, which approximately follows the law of $1/\cos\theta$. However, when DI = 0, the SEU cross-sections of DFF0 and DFF1 at 60° tilt present orders of magnitude differences to the vertical irradiation, which may be related to the parasitic amplification effect caused by the tilt incidence.

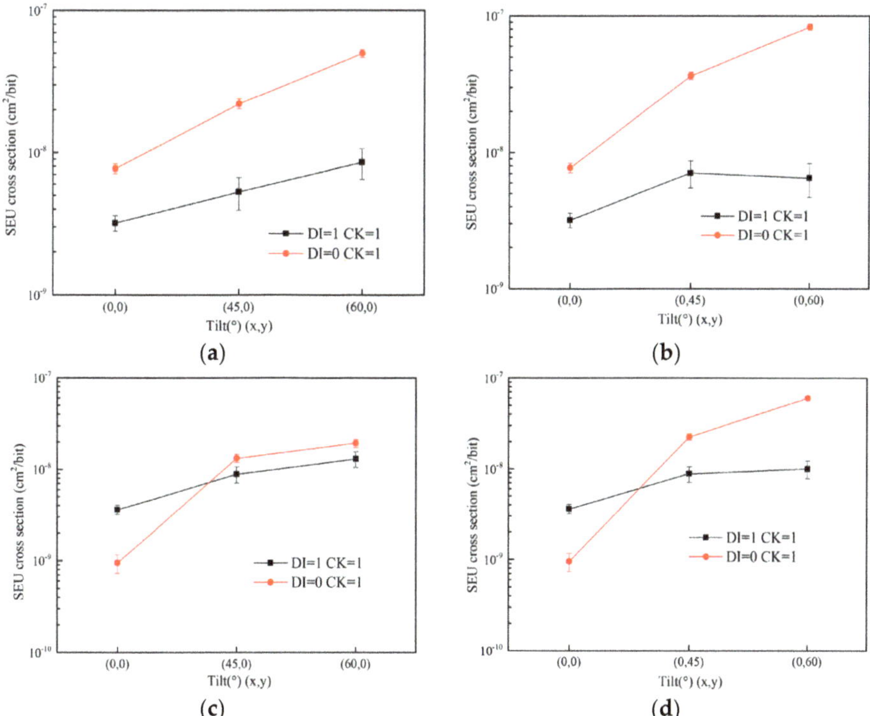

Figure 5. *Cont.*

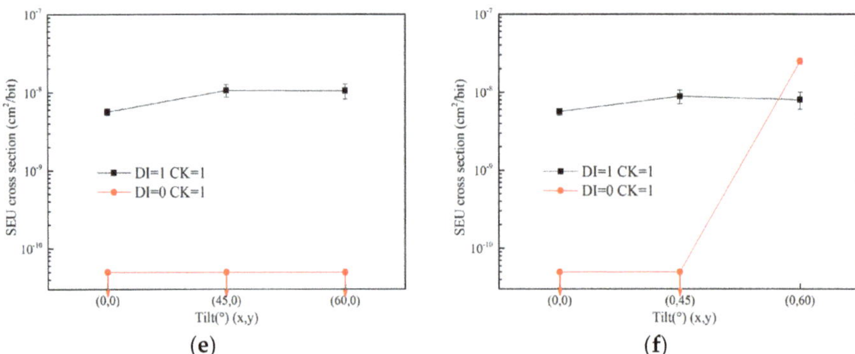

Figure 5. SEU cross-sections of DFFs vs. tilt angles: (**a**) X-direction for DFF0; (**b**) Y-direction for DFF0; (**c**) X-direction for DFF1; (**d**) Y-direction for DFF1; (**e**) X-direction for DFF2; (**f**) Y-direction for DFF2.

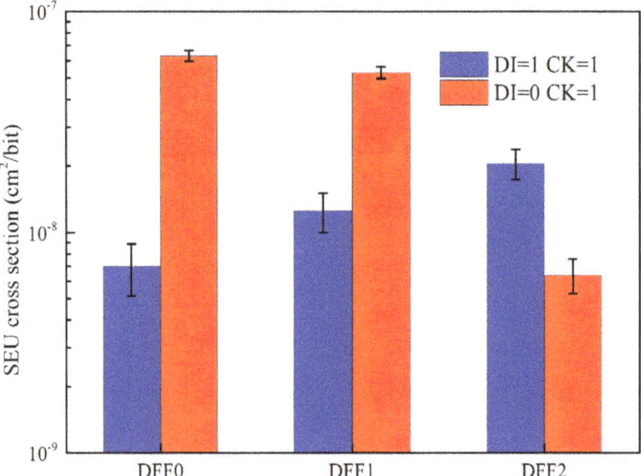

Figure 6. SEU cross-sections of the three DFFs (tilt angle: X = Y = 45°).

The joint change of tilt angles was achieved in the irradiation experiments, and the results are shown in Figure 6. It is clear that the SEU sensitivities of DFFs depend on the data patterns. For DFF0 and DFF1, the SEU cross-sections of full 0 data were higher than that of full 1 data, whereas the data pattern dependency for DFF2 was different. Comparing the three DFFs, the SEU cross-sections for full 0 data decreased from DFF0 to DFF2, while the SEU cross-sections for full 1 data increased from DFF0 to DFF2. Interestingly, the simultaneous variations of along-cell and cross-cell tilt incidences had higher SEU cross-sections than the single 45° tilt incidence for all of the DFFs.

More detailed comparisons for different tilt angles and DFF chains are provided in Figure 7a,b. For full 0 data, the steady decreases of SEU cross-sections for DFF0-2 were measured, which is also consistent with the variation of the along-cell and cross-cell tilt irradiation, indicating that the well isolation was effective for full 0 data. However, for the full 1 data, the mechanisms for the SEU mitigation of well isolation are more complicated, especially for the large 60° tilt. Compared with the results in Figure 7a,b, it was found that the radiation tolerances of full 1 data are much better than the full 0 data for all of the test DFF chains, which is due to the asymmetric structure of the master latch and slave latch in DFF.

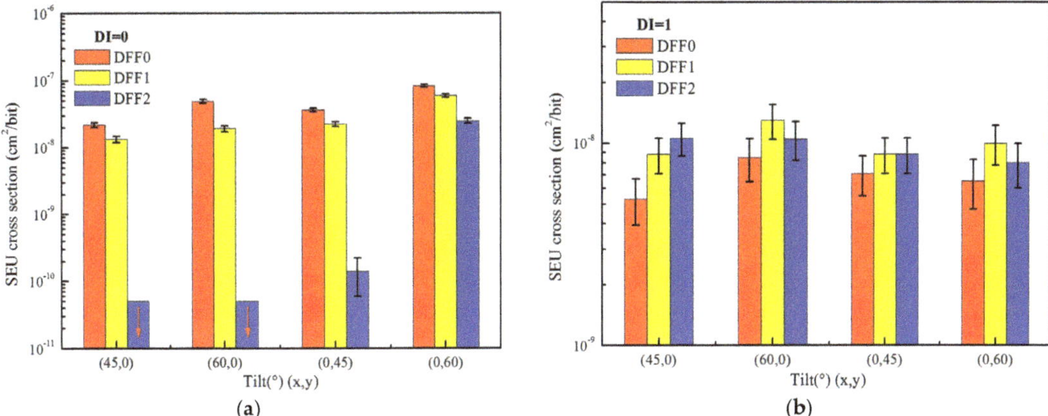

Figure 7. Comparison of SEU cross-sections under different tilt angles for (**a**) DI = 0 and (**b**) DI = 1.

4. Discussion: Hardness Assurances and Failure Analyses

The basic heavy-ion characterizations for the regular bulk planar DFFs and deep-submicron Partially Depleted Silicon on Insulator (PDSOI) DFFs are complete, and the SEU results are reported in references [1–11]. The SEU cross-sections of conventional DFF decrease with the decrease of the technology nodes, whereas the SEU results for the DICE DFF are not consistent with the tendency for feature size shrinking [1,2,7–9]. To minimize the area of the layout, the saturated SEU cross-section of our DFF0 was nearly in the same order of magnitude as SEU cross-sections for the other DICE DFFs provided in previous work [1,2,7–11]. However, the well-isolated DFF1 and DFF2 displayed the same principle of SEU cross-sections only for the condition of full 1 data. For the full 0 data, the hardening effectiveness was much improved. In addition, the SEU sensitivities of DFF manufactured by the 130 nm bulk planar, 130 nm SOI technology, and 22 nm SOI technology have been investigated [6,10]. An approximately two orders of magnitude improvement of SOI DFF in mitigation of SEU cross-sections was verified, indicating that the SOI technology seems to have advantages to further decrease the SEU sensitivities of the nanoscale DICE DFF, and the physical separation of adjacent devices may lead to more enhancements of SEU tolerance for the nanoscale SOI process than the bulk planar process [6,17–19].

For space applications of high-performance electronic systems, the SEU sensitivities of advanced 28 nm technology must be known. Thus, the SEU sensitivities for the 28 nm radiation hardened DFFs are characterized and discussed in Section 3. It was found that the condition of CK = 1 dominated the calculated SEU cross-sections for all of the hardened DFFs, which is due to the approximate sensitive volumes that the DICE DFF cells have. Besides, it is known that the nanoscale devices have limited SEU critical charges, indicating that the charge sharing phenomenon for bulk planar technology can affect the DICE circuits with sufficient charge deposition in coupled SVs. Moreover, it should be noted that the ~3 μm well isolation in 28 nm planar technologies still cannot fully prevent the high-LET heavy ions induced SEUs. Hence, the upsets occurred in these small DFF cells should be fully evaluated before considering whether they are acceptable for space application for a certain mission. Furthermore, though the well isolation applied in DICE DFFs can reduce the SEU rates, it has limited effectiveness due to the lack of well contacts, leading the ionized charges to diffuse and affect more transistors. Therefore, simple well isolation is not a good option for DICE DFFs to further mitigate SEU sensitivities, and in the case of high SEU rates, the well contact seems essential to further mitigate the SEU cross-sections, especially for the layouts with a large spacing of well isolations.

Another interesting phenomenon we observed is that the joint change of tilts can further increase the SEU cross-sections of the hardened circuits, and the mechanisms of

SEU sensitivities for the hardened circuits are related to the actual projective spaces and spaces of SV pairs under different heavy-ion irradiation conditions, as shown in Figure 8. The direction of irradiation with the tilt of X = Y = 45° shows shorter spacing of the transistors than the individual tilt of X = 45° or Y = 45°, meaning that more serious SEU sensitivity will be observed under the X = Y = 45° conditions. It is clear that the tilt incidence of high-LET heavy ions can have a more serious influence, especially for the redundancy hardened circuits. Therefore, considering the 4π-distributed high-energy heavy ions in space environments, the evaluation of SEU for both the vertical and tilt irradiations of high-LET ions is necessary to eliminate the overestimation of SEU tolerance and guarantee the in-orbit safety of spacecraft in harsh radiation environments.

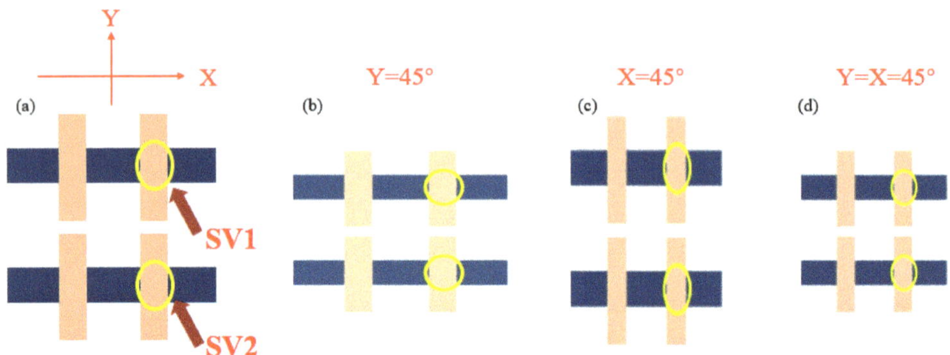

Figure 8. Diagrams for the equivalent projective shapes and spaces of SV pairs under diverse heavy-ion tilt incidences. (**a**) vertical irradiation, (**b**) Y = 45° tilt, (**c**) X = 45° tilt, and (**d**) X = Y = 45° tilt.

5. Conclusions

In this paper, the SEU performances of different DICE DFFs fabricated with an advanced 28 nm planar technology are presented. The proportions of SEU cross-sections for the different irradiation conditions are distinguished and classified. The different clock dependency is related to the unbalanced structure of master latch and slave latch in DFFs, which can be explained by a greater than six times difference in the number of upset nodes. The DICE DFF2 is SEU immune for the full 0 data until tilt angles over 45°, whereas the 60° tilt incidences make the ~3-µm well isolation less effective. The abundant testing stresses combined with diverse layout structures indicate that the SEU immunity is hard to achieve for the 28 nm planar technology. Though the area consumption of the well isolation is non-negligible, the improvements of SEU tolerance are not obvious. In addition, the joint changes of tilts (X = Y = 45°) improve the SEU sensitivity of the hardened DFFs, which needs full consideration for the space application of hardening circuits due to the existence of long-range high-LET heavy ions in space environments. The heavy-ion evaluations are useful for the related integrated circuits and provide data to support the radiation hardening design of 28 nm technology.

Author Contributions: Conceptualization, Y.C., Z.W., Z.H. and Y.F.; methodology, J.C., C.C. and B.L.; software, Y.C.; validation, Z.W. and Z.H.; formal analysis, Y.F.; investigation, J.C.; resources, Y.C.; data curation, C.C.; writing—original draft preparation, Y.C. and C.C.; writing—review and editing, Z.W.; visualization, C.C.; supervision, B.L.; project administration, Y.C.; funding acquisition, Y.C. All authors have read and agreed to the published version of the manuscript.

Funding: This research was funded by the National Natural Science Foundation of China, grant number 61704192, 62004221, and 62174180; the HIRFL, grant number JIZR20GY002; the foundation of NUDT, grant number ZK20-25.

Conflicts of Interest: The authors declare no conflict of interest.

References

1. Amusan, O.A.; Massengill, L.W.; Baze, M.P. Single Event Upsets in Deep-Submicrometer Technologies Due to Charge Sharing. *IEEE Trans. Device Mater. Reliab.* **2008**, *8*, 582–589. [CrossRef]
2. Gaspard, N.; Jagannathan, S.; Diggins, Z. Angled flip-flop single-event cross sections for submicron bulk CMOS technologies. In Proceedings of the European Conference on Radiation and Its Effects on Components and Systems (RADECS), Oxford, UK, 23–27 September 2013.
3. Lin, T.; Chong, K.S.; Shu, W. Experimental Investigation into Radiation-Hardening-By-Design (RHBD) Flip-Flop Designs in a 65 nm CMOS Process. In Proceedings of the IEEE International Symposium on Circuits and Systems (ISCAS), Montreal, QC, Canada, 22–25 May 2016.
4. Diggins, Z.J.; Gaspard, N.J.; Mahatme, N.N. Scalability of Capacitive Hardening for Flip-Flops in Advanced Technology Nodes. *IEEE Trans. Nucl. Sci.* **2013**, *60*, 4394–4398. [CrossRef]
5. Quinn, R.C.; Kauppila, J.S.; Loveless, T.D. Heavy Ion SEU Test Data for 32 nm SOI Flip-Flops. In Proceedings of the IEEE Radiation Effects Data Workshop (REDW), Boston, MA, USA, 13–17 July 2015.
6. Cai, C.; Liu, T.Q.; Zhao, P.X. Multiple Layout-Hardening Comparison of SEU-Mitigated Filp-Flops in 22-nm UTBB FD-SOI Technology. *IEEE Trans. Nucl. Sci.* **2020**, *67*, 374–381. [CrossRef]
7. Wang, H.; Dai, X.; Ibrahim, Y.M. A layout-based rad-hard DICE flip-flop design. *J. Electron. Test.* **2019**, *35*, 111–117. [CrossRef]
8. Jagannathan, S.; Loveless, T.D.; Bhuva, B.L. Single-Event Tolerant Flip-Flop Design in 40-nm Bulk CMOS Technology. *IEEE Trans. Nucl. Sci.* **2011**, *58*, 3033–3037. [CrossRef]
9. Lilja, K.; Bounasser, M.; Wen, S.J. Single-Event Performance and Layout Optimization of Flip-Flops in a 28-nm Bulk Technology. *IEEE Trans. Nucl. Sci.* **2013**, *60*, 2782–2788. [CrossRef]
10. Zhang, L.; Xu, J.; Fan, S. Single Event Upset Sensitivity of D-Flip Flop: Comparison of PDSOI with Bulk Si at 130 nm Technology Node. *IEEE Trans. Nucl. Sci.* **2017**, *64*, 683–688. [CrossRef]
11. Cai, C.; Fan, X.; Liu, J. Heavy-Ion Induced Single Event Upsets in Advanced 65 nm Radiation Hardened FPGAs. *Electronics* **2019**, *8*, 323. [CrossRef]
12. Hansen, D.L.; Miller, E.J.; Kleinosowski, A. Clock, Flip-Flop, and Combinatorial Logic Contributions to the SEU Cross Section in 90 nm ASIC Technology. *IEEE Trans. Nucl. Sci.* **2009**, *56*, 3542–3550. [CrossRef]
13. Warren, K.M.; Sierawski, B.D.; Reed, R.A. Monte-Carlo Based On-Orbit Single Event Upset Rate Prediction for a Radiation Hardened by Design Latch. *IEEE Trans. Nucl. Sci.* **2007**, *54*, 2419–2425. [CrossRef]
14. Chatterjee, I.; Narasimham, B.; Mahatme, N.N. Impact of Technology Scaling on SRAM Soft Error Rates. *IEEE Trans. Nucl. Sci.* **2014**, *61*, 3512–3518. [CrossRef]
15. Chatterjee, I.; Narasimham, B.; Mahatme, N.N. Single-Event Charge Collection and Upset in 40-nm Dual- and Triple-Well Bulk CMOS SRAMs. *IEEE Trans. Nucl. Sci.* **2011**, *58*, 2761–2767. [CrossRef]
16. Zhang, Z.; Liu, J.; Hou, M. Investigation of Threshold Ion Range for Accurate Single Event Upset Measurements in Both SOI and Bulk Technologies. *IEEE Trans. Nucl. Sci.* **2014**, *61*, 1459–1467. [CrossRef]
17. Ferlet-Cavrois, V.; Massengill, L.W.; Gouker, P. Single Event Transients in Digital CMOS—A Review. *IEEE Trans. Nucl. Sci.* **2013**, *60*, 1767–1790. [CrossRef]
18. Rodbell, K.P.; Heidel, D.F.; Pellish, J.A. 32 and 45 nm Radiation-Hardened-by-Design (RHBD) SOI Latches. *IEEE Trans. Nucl. Sci.* **2011**, *58*, 2702–2710. [CrossRef]
19. Heidel, D.F.; Marshall, P.W.; Pellish, J.A. Single-Event Upsets and Multiple-Bit Upsets on a 45 nm SOI SRAM. *IEEE Trans. Nucl. Sci.* **2009**, *56*, 3499–3504.

 *electronics*

MDPI

Article

Novel Radiation-Hardened High-Speed DFF Design Based on Redundant Filter and Typical Application Analysis

Shuang Jiang [1,2,*], Shibin Liu [1], Hongchao Zheng [2], Liang Wang [2] and Tongde Li [2]

1 The School of Electronics and Information, Northwestern Polytechnical University, Xi'an 710072, China; liushibin@nwpu.edu.cn
2 Beijing Microelectronics Technology Institute, Beijing 100076, China; hongchao_zheng@163.com (H.Z.); wangliang150200@163.com (L.W.); ltdeam@163.com (T.L.)
* Correspondence: jiangsh_0425@163.com

Abstract: A cell-level radiation hardening by design (RHBD) method based on commercial processes of single event transient (SET) and single event upset (SEU) is proposed in this paper, in which new radiation-hardened D-type flip-flops (DFFs) are designed. An application-specific integrated circuit (ASIC) of a million gates level is developed based on DFFs, and SEU and single event functional interruption (SEFI) heavy-ion radiation tests are carried out. The experimental results show that the new DFF SEU ability is increased by 63 times compared with the DICE-designed DFF, and is three orders of magnitude higher than the redundantly designed DFF. The SEFI ability of the ASIC designed by the new DFF is 2.6 times higher than the circuit hardened by the TMR design.

Keywords: D-type flip-flop; single event transient; single event upset; radiation hardened

Citation: Jiang, S.; Liu, S.; Zheng, H.; Wang, L.; Li, T. Novel Radiation-Hardened High-Speed DFF Design Based on Redundant Filter and Typical Application Analysis. *Electronics* **2022**, *11*, 1302. https://doi.org/10.3390/electronics11091302

Academic Editor: Paul Leroux

Received: 23 March 2022
Accepted: 18 April 2022
Published: 20 April 2022

Publisher's Note: MDPI stays neutral with regard to jurisdictional claims in published maps and institutional affiliations.

1. Introduction

Heavy ions and high energy protons in space can produce single event effects (SEE) on semiconductor devices, while single event soft errors in digital integrated circuits have become the main cause of failure of space vehicles. First, SEU is more sensitive in a small size process, which makes the multi node upset (MNU) occur more easily in the adjacent device due to the charge sharing [1]. Second, with the increasing working frequency of the circuit, SET is more likely to be captured by sequential units and be converted to SEU [2]. Third, the single continuous error (SCU) is easier to generate due to the clock tree affected by SET as the circuit scale increases [3], which will lead to the probability of single event soft errors being greatly increased in digital integrated circuits.

The RHBD method based on the commercial process has the advantages of not needing to modify the process parameters, having a low cost of tape-out, and having a good hardened performance, and has been widely used in the development of aerospace integrated circuits. The guard ring structures method is commonly used in the layout level of anti-SET [4]. The single event charge sharing can be suppressed by inserting minority carrier guard rings between the adjacent drains. The cell-level anti-SET and anti-SEU methods of DFF include triple modular redundancy (TMR) [5], C cell [6], dual interlocked cell (DICE) [7], and error correction code (ECC) [8], which improve the anti-SEE performance at the expense of area or time.

To improve the anti-SEU capability of DFF, considering SETs from buffers or logic gates inside the cell may cause DFF errors. A new DFF IP cell that is resistant to SET, SEU, and MNU, is designed by optimizing the filter of the DFF input to reduce the influence of SET [1,3]. In addition, two types of DFFs with different reinforcement levels are used for the circuit design, which can save layout area overhead and ensure a high SEU resistance by rationally screening and replacing DFFs on non-critical paths. Finally, the design, tape-out, and radiation tests of the verification circuit ASIC based on the 0.18 μm commercial standard CMOS process are completed through the above methods.

2. DFF Design

A novel DFF structure based on redundant delay filter (RDF) and dual DICE is designed, as shown in Figure 1 [3], and is called RDD-DFF. DICE is used to improve the SEU threshold, while RDF is used to filter out the external SET. The two outputs in our design are independent, so the SET generated by a single output will not affect the dual-mode redundant DICE latch. In this way, the source of the SET can be eliminated maximally.

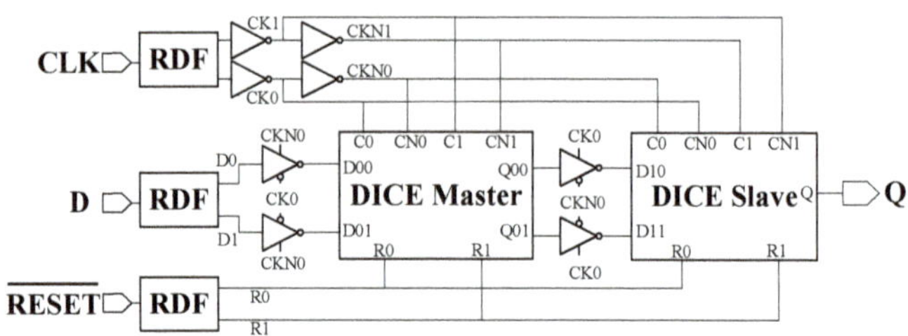

Figure 1. The RDD-DFF structure.

2.1. SET Simulation

The SET simulations were performed on Sentaurus TCAD, applying a Fermi−Dirac statistics and hydrodynamic model, taking into account the effects of doping, electric field, carrier−carrier scattering, and interface scattering on mobility, as well as the effect of band-gap narrowing, while the temperature was set at 25 °C. The simulation results of Bi−923.2MeV and Cl−158MeV are shown in Figure 2a,b, while the arrow direction is the vertical 90−degree incident direction of the heavy ions. The SET horizontal track widths under irradiation can be obtained at about 1 μm. The electrostatic potential distribution is shown in Figure 2c. The SET sensitive area generated by the heavy ions is mainly on the drain region of the device in the off state. At this time, the drain PN junction is at a reverse bias, and a depletion region is formed, which is the main region for collecting charges. The holes induced by radiation are collected at the drain, resulting in the potential of the drain increasing.

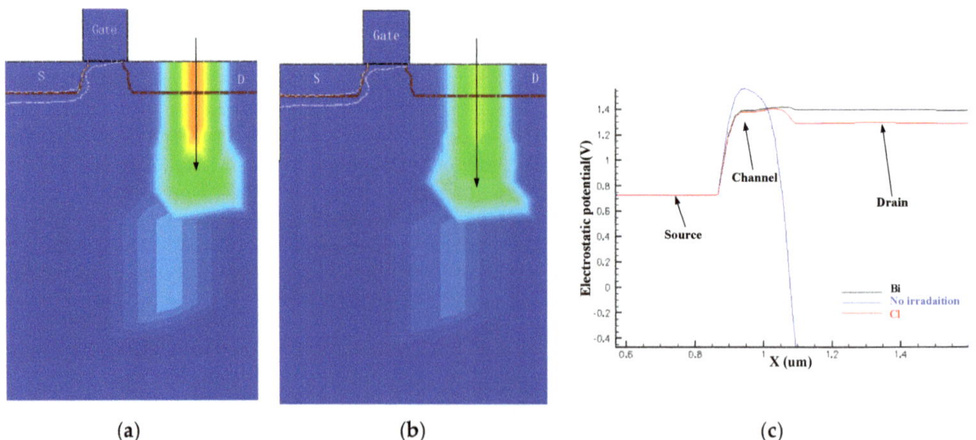

Figure 2. The TCAD simulation results of SET: (**a**) Bi−923.2MeV; (**b**) Cl−158MeV; (**c**) electrostatic potential distribution.

2.2. Anti-SET Design

Charge sharing comes from diffusion effects in NMOS, which is the main course of SET on the double-well process, while for PMOS, charge sharing mainly comes from the bipolar amplification. The guard rings and contact holes are designed to speed up the collection of interfering charges to reduce the SET pulse width. In addition, it is also necessary to strengthen the charge sharing of transistors with different polarities. With the guidance of the SET width obtained by simulation, the isolation of the sensitive node pairs is realized by introducing another complementary well between the same phase nodes. The distances between the same phase nodes are designed to be greater than 2 μm, as shown in Figure 3. In this way, the charge between the same polarity transistors can be eliminated, and the charge deposited on the sensitive nodes is reduced.

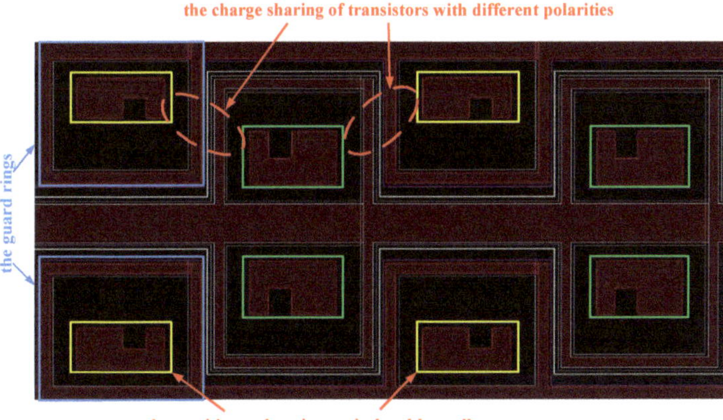

Figure 3. The hardened layout using charge sharing.

The RDF structure is designed as shown in Figure 4 [3]. The filtering delay threshold is set to be adjustable. The appropriate filtering threshold can be set according to the SET circuit test results, while a corresponding time overhead is required.

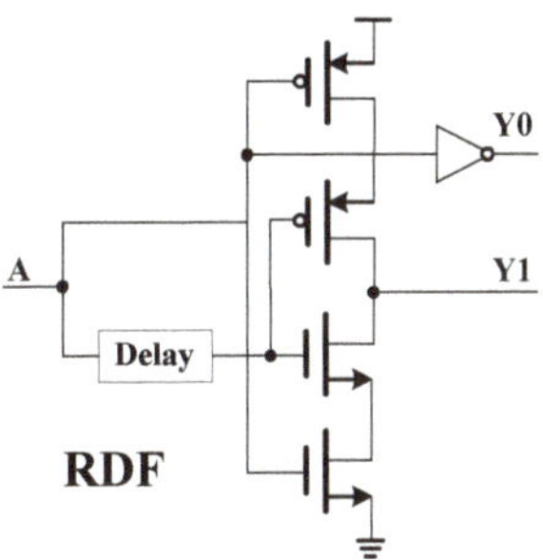

Figure 4. RDF structure.

2.3. Anti-SEU Design

The SEU-sensitive node pairs in the DICE memory cell are shown in Figure 5. The state of node A is flipped from 1 to 0 when the heavy ion is irradiated on the reverse-biased NMOS drain region (n1), so that NMOS-M7 is turned off and PMOS-M4 is turned on, which makes the state of node B flip from 0 to 1. Then, PMOS-M6 will turn off while nodes C and D are both in a floating state. The DICE cell will finally flip if both nodes are simultaneously affected by the charge sharing effect or oblique angle injection. Therefore, all sensitive node pairs should be properly laid out during the layout design stage.

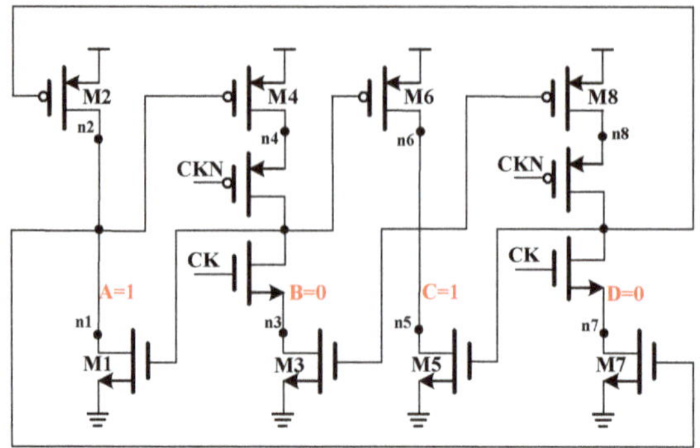

Figure 5. The sensitive node analysis of the DICE cell.

The team has proposed an error quenching double DICE (EQDD) method using the layout crossing of two DICE cells to improve the SEU ability of DICE [1], through which the SEU LET threshold could effectively be improved. The charge sharing between the non-sensitive nodes in one DICE cell can be used to reduce the SET error through a quenching effect, while the distance between the sensitive nodes within adjacent DICEs can be set to more than 6 μm. In this way, the sensitive node pairs can be separated without losing the area cost through the layout crossing design of two DICE cells, as shown in Figure 6.

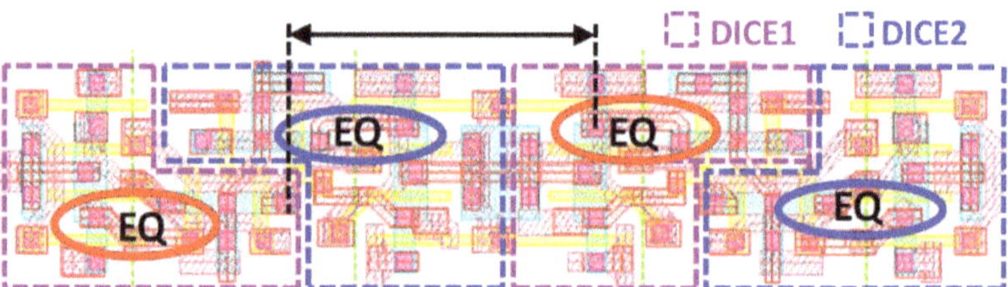

Figure 6. Layout of the EQDD method.

2.4. Selective SEU Design

The RDD-DFF design needs an extra delay and area, although it has both a better SEU and SET performance. At the VLSI level, DFFs can be chosen at different anti-SEE levels according to the timing constraints. The DFF on the non-critical path can be replaced with RDD-DFF to improve the SEE capability under the premise that the overall circuit delay will not be increased after replacement.

All path delay values can be comprehensively enumerated with the help of the DC tool to import the circuit netlist, through which the critical path can be formed by counting the maximum path delay. The improvement of the circuit SEE can be obtained at the cost of a minimal area overhead, with the help of a program written in C# (as shown in Figure 7) to identify alternative DFFs.

```
//Circuit_DC: Circuit Delay Information File
//DFF_Compilations[]: DFF array of Name&Delay&ProTime
Wihle(!Circuit_DC.EndOfStream)
  {
    If (Circuit_DC.ReadLine().Contain( " DFF " ))
    DFF_Compilations.add(DFF_Name, DFF_Delay,DFF_ProTime);
  }
DFF_Max_Delay=MAX(DFF_Compilations.Delay);
For (i=0;i<DFF_Compilations.Size();i++)
  {
    Deta_t=RDD_DFF.ProTime-DFF_Compilations[i].ProTime;
    if (DFF_Max_Delay>(DFF_Compilations[i].delay+Deta_t))
      Replace(DFF_Compilations[i].name,RDD_DFF);
  }
```

Figure 7. The algorithm to identify alternative DFFs.

3. Radiation Tests

An ARINC 659 bus protocol control circuit (ASIC) is designed using the RDD-DFF cell with the above-mentioned RHBD methods at 0.18 μm through the CMOS process. The circuit scale is about 1.65 million gates with 26,725 DFFs, which has the highest operating frequency at 144 MHz. The appearance of the chip is shown in Figure 8.

Figure 8. The chip appearance.

An SEU test system based on a scan chain is developed to verify the RDD-DFF SEE performance. The test clock frequency is set at 1 MHz and the internal DFFs are in the dynamic mode so as to read back the "01" code stream. A typical functional SEFI test system is also developed to evaluate the SEU performance of the circuit. FPGA is used to input the same excitation vector to both the device being tested and the comparison device, while the operating frequency is set at 144 MHz. The output results of the device being tested and the comparison device are collected separately by FPGA, while the real-time comparison result is used to determine whether a single event function interruption or error occurred. The SEE experiments are carried out on the HIRFL cyclotron accelerator in Lanzhou and the HI-13 tandem accelerator in Beijing, respectively. The heavy ions are shown in Table 1. The test site picture is shown in Figure 9.

Table 1. The heavy ion parameters.

Ion	Energy (MeV)	LET (MeV × cm²/mg)	Range (μm)	Accelerator
Cl	158	13.1	51.1	
Ti	169	21.8	37.9	HI-13
Ge	205	37.3	30	
Bi	923.2	99.8	53.7	HIRFL

Figure 9. SEE test on HIRFL.

4. Results and Discussion

The SEU results of RDD-DFF obtained based on the SCAN method are shown with the red line in Figure 10, while the SEU results of the DICE-DFF that did not use the EQDD method before are shown with the blue line for comparison. The other results of the 0.18 μm CMOS DFF (DFF-R) designed using the RHBD method of two redundant storage node topologies proposed in [9] are shown with the black line. The SEU saturated cross-section is obtained by fitting and drawing the Weibull curve, while the SEU LET threshold value can be taken as corresponding to 10% of the saturated cross-section. The SEU on-orbit error rates normalized to each bit is shown in Table 2, which are obtained using the RPP model in the radiation environment of the Adams 90% maximum bad case and the 3-mm equivalent Al shielding.

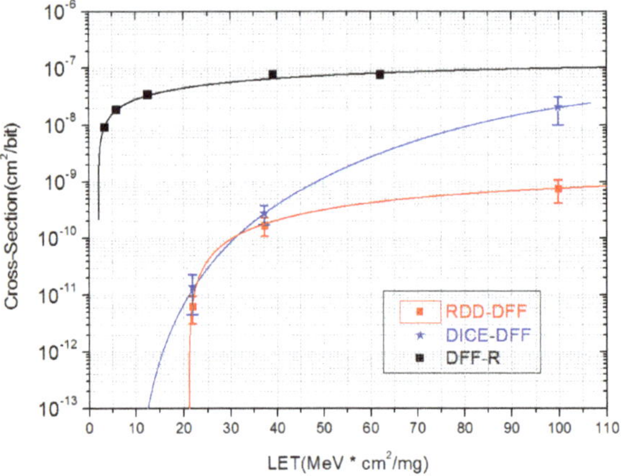

Figure 10. The SEU Weibull curve.

Table 2. The heavy ion parameters.

Types	Saturation Cross Section (cm²/bit)	LET Threshold (10%) (MeV × cm²/mg)	SEU On-Orbit Error Rate (/bit/day)
RDD-DFF	7.48×10^{-10}	28.4	5.50×10^{-10}
DICE-DFF	2.05×10^{-8}	35.0	3.51×10^{-8}
DFF-R	7.90×10^{-8}	3.1	3.87×10^{-6}

The results show that the SEU of RDD-DFF is similar to the DFF using ordinary DICE under the condition of small LET ions, which can be due to the small LET ions having a small charge sharing radius, which is difficult to occur through SEU in DICE. Thereof, the probability of SEU occurrence in DICE using EQDD is equivalent to the traditional DICE, while the SEU of DFF is mainly derived from the SET occurring in RDF and CLK. However, the SEU cross-section is reduced by about 27 times under large LET ions by using the EQDD method, which means it is more effective against SEU than the traditional DICE. The SEU error rate of RDD-DFF is 63 times better than DICE-DFF and four orders of magnitude better than DFF-R, which indicates that DICE plays an important role in SEU, as well as RDF in SET.

To compare the SEFI performance, the SEFI Weibull curves of the designed ASIC, the DSP circuit SMV320C6701 at 0.18 μm process in [10], and the DSP circuit RTAX4000D at 0.15 μm process in [11] are drawn in Figure 11, while the SEFI index is calculated as shown in Table 3.

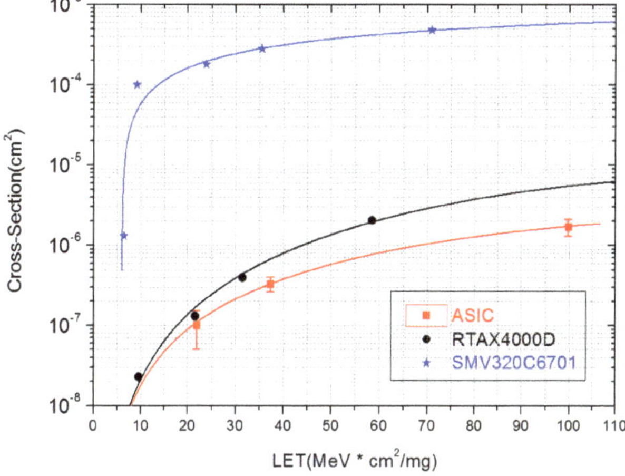

Figure 11. The SEFI Weibull curve.

Table 3. The heavy ion parameters.

Types	Frequency	Saturation Cross Section (cm^2/bit)	LET Threshold (10%) (MeV $\times$ cm^2/mg)	SEFI On-Orbit Error Rate (/bit/day)
ASIC	144 MHz	1.7×10^{-6}	27.2	8.12×10^{-7}
RTAX4000D	120 MHz	5.4×10^{-6}	34.0	2.11×10^{-6}
SMV320C6701	140 MHz	4.8×10^{-4}	9.4	3.57×10^{-3}

As the key status and data registers in complex circuits such as ASIC and DSP are generally DFF cells, they are likely to cause a disorder of the state machine of DUT and lead to SEFI, while SEU occurs on these registers. The SEFI probability of ASIC is 2.6 times smaller than that of RTAX4000D due to the better SEU performance of DICE-DFF. The R-cell in RTAX4000D is a TMR design, which has a better resistance to SEU, while the filter is used to improve the SET performance. In contrast, the ASIC in this paper adopt hardened-design of DFF cells instead of TMR strategy, which can effectively reduce the layout area, power consumption, and the operating fre-quency performance.

The SEFI probability of ASIC is three orders of magnitude smaller than SMV320C6701, while the SEU results of the cache, memory, and other storage cells in SMV320C6701 show that no effective SEU hardening design is carried out at a cell level. In contrast,

ASIC does not reduce the operating frequency, which is the same advantage as using the RHBD method.

5. Conclusions

In the field of aerospace, more attention has been paid to the cost of chips, which makes obtaining reliable radiation resistance and a high performance with the smallest area cost an eternal topic. In this paper, the SEU performance is improved 63 times through designing a new hardening circuit structure and layout innovation, as well as through using selective optimization methods of DFF. The traditional cell-level hardening design of DFF is improved, while maintaining no additional increase in area. The SEFI performance of ASIC hardening by the cell is better than using the TMR method by about 2.6 times, which indicates that cell-level hardening design is one of the most cost-effective ways to design aerospace integrated circuits.

Author Contributions: Conceptualization, S.J. and S.L.; methodology, S.J.; software, H.Z.; validation, S.J., L.W. and T.L.; formal analysis, S.J.; investigation, T.L.; resources, H.Z.; data curation, L.W.; writ-ing—original draft preparation, S.J.; writing—review and editing, S.L. and H.Z.; visualization, H.Z.; su-pervision, S.L.; project administration, S.L.; funding acquisition, S.J. All authors have read and agreed to the published version of the manuscript.

Funding: This research received no external funding.

Institutional Review Board Statement: Not applicable.

Informed Consent Statement: Not applicable.

Data Availability Statement: Not applicable.

Conflicts of Interest: The authors declare no conflict of interest.

References

1. Zhao, Y.F.; Wang, L.; Yue, S.G. Single Event Effect and its Hardening Technique in Nano-scale CMOS Integrated Circuits. *Acta Electron. Sin.* **2018**, *46*, 2511–2518.
2. Chia-Hsiang, C.; Phil, K. Characterization of Heavy-Ion-Induced Single-Event Effects in 65 nm Bulk CMOS ASIC Test Chips. *IEEE Trans. Nucl. Sci.* **2014**, *61*, 2694–2701.
3. Zhao, Y.; Wang, L.; Yue, S.; Wang, D.; Zhao, X.; Sun, Y.; Li, D.; Wang, F.; Yang, X.; Zheng, H.; et al. SEU and SET of 65 Bulk CMOS Flip-flops and Their Implications for RHBD. *IEEE Trans. Nucl. Sci.* **2015**, *62*, 2666–2672. [CrossRef]
4. Narasimham, B.; Gambles, J.W.; Shuler, R.L. Quantifying the Effect of Guard Rings and Guard Drains in Mitigating Charge Collection and Charge Spread. *IEEE Trans. Nucl. Sci.* **2008**, *55*, 3456–3460. [CrossRef]
5. Blum, D.R.; Delgado-Frias, J.G. Schemes for eliminating transient-width clock overhead from SET-tolerant memory-based systems. *IEEE Trans. Nucl. Sci.* **2006**, *53*, 1564–1573. [CrossRef]
6. Nan, H.Q.; Ken, C. High Performance, Low Cost and Robust Soft Error Tolerant Latch Designs for Nanoscale CMOS Technology. *IEEE Trans. Circ. Syst.* **2012**, *59*, 1445–1457. [CrossRef]
7. Calin, T.; Nicolaidis, M.; Velazco, R. Upset hardened memory design for submicron CMOS technology. *IEEE Trans. Nucl. Sci.* **1996**, *43*, 2874–2878. [CrossRef]
8. Saiz-Adalid, L.J.; Reviriego, P.; Gil, P. MCU Tolerance in SRAMs Through Low-Redundancy Triple Adjacent Error Correction. *IEEE Trans. VLSI Syst.* **2015**, *23*, 2332–2336. [CrossRef]
9. Gaspard, N.J.; Jagannathan, S. Technology Scaling Comparison of Flip-Flop Heavy-Ion Single-Event Upset Cross Sections. *IEEE Trans. Nucl. Sci.* **2013**, *60*, 4368–4373. [CrossRef]
10. Joshi, R.; Daniels, R. Radiation Hardness Evaluation of a Class V 32-Bit Floating-Point Digital Signal Processor. In Proceedings of the IEEE Radiation Effects Data Workshop, Seattle, WA, USA, 11–15 July 2005.
11. Sana, R.; Paul, L. SEE Characterization of the New RTAX-DSP (RTAX-D) Antifuse-Based FPGA. *IEEE Trans. Nucl. Sci.* **2010**, *57*, 3537–3546.

MDPI

Article

A Fully Polarity-Aware Double-Node-Upset-Resilient Latch Design

Jung-Jin Park [1], Young-Min Kang [2], Geon-Hak Kim [2], Ik-Joon Chang [2] and Jinsang Kim [2,*]

[1] Department of Electrical and Computer Engineering, University of Washington, Seattle, WA 98195, USA
[2] Department of Electronic Engineering, Kyung Hee University, Yongin 17104, Korea
* Correspondence: jskim27@khu.ac.kr; Tel.: +82-31-201-2996

Abstract: Due to aggressive scaling down, multiple-node-upset hardened design has become a major concern regarding radiation hardening. The proposed latch overcomes the architecture and performance limitations of state-of-the-art double-node-upset (DNU)-resilient latches. A novel stacked latch element is developed with multiple thresholds, regular architecture, increased number of single-event upset (SEU)-insensitive nodes, low power dissipation, and high robustness. The radiation-aware layout considering layout-level issues is also proposed. Compared with state-of-the-art DNU-resilient latches, simulation results show that the proposed latch exhibits up to 92% delay and 80% power reduction in data activity ratio (DAR) of 100%. The radiation simulation using the dual-double exponential current source model shows that the proposed latch has the strongest radiation-hardening capability among the other DNU-resilient latches.

Keywords: double-node upset (DNU); radiation-hardened latch; radiation hardening by design (RHBD); single event upset polarity; single-node upset (SNU); soft error; static random-access memory (SRAM)

Citation: Park, J.-J.; Kang, Y.-M.; Kim, G.-H.; Chang, I.-J.; Kim, J. A Fully Polarity-Aware Double-Node-Upset-Resilient Latch Design. *Electronics* **2022**, *11*, 2465. https://doi.org/10.3390/electronics11152465

Academic Editors: Antonio Di Bartolomeo and Paul Leroux

Received: 27 June 2022
Accepted: 2 August 2022
Published: 8 August 2022

Publisher's Note: MDPI stays neutral with regard to jurisdictional claims in published maps and institutional affiliations.

1. Introduction

Recent advances in scaling down technology have led to a substantial decrease in supply voltage and node capacitance, resulting in a smaller amount of critical charge, the minimum charge necessary to maintain a logic state. This implies that a soft error caused by single-event upset (SEU) may occur not only in harsh radiation environments at high altitudes but also at the terrestrial level [1]. Considering the decreasing feature size in the current nanometer technology, multiple-node-upset (MNU) is more likely to occur because of charge sharing among nodes [2]. An empirical study reported that 40-nm flip-flops implemented using single-node upset (SNU)-hardened dual-interlocked memory cell (DICE) [3] do not entirely prevent the soft error and exhibit only approximately 30% better cross sections compared with non-radiation hardened flip-flops [4]. Thus, recent research on radiation hardening by design (RHBD) has focused on the double-node-upset (DNU)-hardened designs, extending the SNU-hardened latches as primitive circuit elements.

The DNU-hardened latch designs can be classified into three circuit elements: the Muller C-element (MCE) [5,6], DICE [7,8], and Schmitt-trigger (ST) cell [9]. The MCE-based DNU latches [5,6] are capable of blocking SEU propagation to adjacent nodes by changing the high-impedance state of the MCE outputs when the input nodes suffer from SEU. The output's floating state may cause state flipping at system level when the system accesses them. Because of charge sharing between the access node and the floating output node, the dynamic voltage ripple can flip the state of the floating node. The DICE-based DNU latches extend the DICE to a delta-like [7] or a donut-like [8] architecture. When new input data are written, they consume more power and require longer write time because all the internal feedback loops should be activated and flipped. Additionally, when an upset occurs, the upset node may cause an adjacent node to transit and be suffered from the short path. The ST-based DNU latch [9] is based on the ST inverters. Therefore, it features a high noise margin, low complexity, and low power dissipation. However, its radiation hardening capability is no longer maintained when there is a ratio issue between two feedback inverters.

Recently, an approach [10] to reduce the number of SEU-sensitive nodes in DNU-hardened latches was proposed. The concept was derived from the upset polarity [11] of a CMOS logic inverter. When an energetic particle hits an off-pMOSFET, only 0-to-1 positive upset occurs at the drain node because the pMOSFET only collects positive charges. In contrast, when an energetic particle hits an off-nMOSFET, only 1-to-0 negative upset occurs at the drain node because the nMOSFET only collects negative charges. Based on this principle, a node avoids a negative upset if pMOSFETs are stacked at the node and avoids a positive upset if nMOSFETs are stacked. However, the polarity-aware latch inevitably suffers from V_{th} drop because the latch should stack the same type of MOSFETs. This V_{th} drop causes high power consumption during the state-holding phase.

This paper proposes a *fully polarity-aware DNU-resilient latch* (FPADRL) featuring up to a reduction of 92% delay and 80% power in data activity ratio (DAR) of 100% over existing DNU-resilient latches. Furthermore, the proposed FPADRL overcomes the limitations in the previous research works: (1) resilience to SEU without the charge sharing issue at the system level; (2) fully polarity-aware latch elements, resulting in the maximum possible number of SEU-insensitive internal nodes; (3) fewer short path cases during the SEU with lower power dissipation; (4) much robust radiation-hardening capability; (5) radiation-aware layout.

The rest of this paper is organized as follows. The proposed DNU-resilient latch is described in Section 2. Next, the simulation and evaluation results are presented in Section 3. Finally, Section 4 concludes this paper.

2. Proposed DNU-Hardened Latch Design

2.1. Overall Structure and Design Idea

Figure 1 shows the structure of the two latch elements in the proposed DNU-resilient latch. As shown in Figure 1a, the pMOSFET-stacked latch with the cross-coupled inverter structure has additional stacked pMOSFETs, whose operations are controlled by other latch nodes. When node X1 stores logic "1", this node is insensitive to SEU due to the error polarity principle [11]. Node X2 is only sensitive to the 0-to-1 transition. Similarly, as shown in Figure 1b, when node X6 initially stores logic "0", this node is insensitive to SEU, and node X7 is only sensitive to the 1-to-0 transition. When a radiation particle strikes the floating nodes A1 or A6, the initial value of these nodes may change. However, the floating nodes are not connected to any inputs; thereby, their upsets do not affect any data nodes Xn ($n = 1, \ldots, 8$). The nodes between the on-transistors, such as A2 and A7, are also insensitive to SEU. Therefore, the proposed four-node latch element has only one single polarity SEU sensitive node, one floating node, and two SEU-insensitive nodes.

Simultaneously, the stacked MOSFETs play an essential role in blocking the SEU error propagation. Unlike the DICE-based latch structures, the proposed latch does not propagate the error in all SNU cases and some DNU cases due to the off-stacked transistors. Moreover, the off-stacked transistors generally block the propagation of the upset on the floating node to the data nodes. The detailed mechanism will be explained in Section 2.3.

Figure 2 shows the schematic of the proposed FPADRL. The FPADRL consists of the stacked latches shown in Figure 1, the access transistor module, and the clocked output. The upper P-stacked and lower N-stacked latch parts (PSLP and NSLP) consists of the stacked pMOSFET latches and the stacked nMOSFET latches, respectively. Thus, only positive upset occurs in the PSLP, and only negative upset occurs in the NSLP. The operations of the stacked MOSFETs are controlled by the internal nodes of the opposite latch part. Given that the two latch parts store the same data in order (i.e., X1 = X5, X2 = X6, X3 = X7, and X4 = X8), the respective node values are fed into the gate inputs of the correspondent stacked transistors to form the DICE-like dual-interlocked loops. Compared with recent research works [5,6,10], the proposed FPADRL has regular latch element architecture, and the number of sensitive nodes is maximally reduced from sixteen to eight.

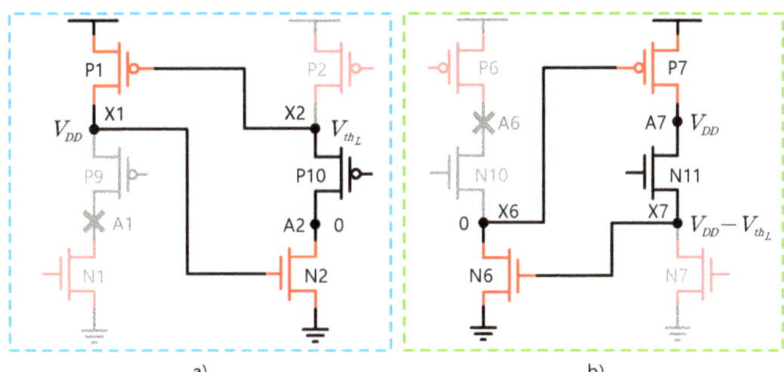

Figure 1. Latch elements: (**a**) pMOSFET-stacked latch, (**b**) nMOSFET-stacked latch.

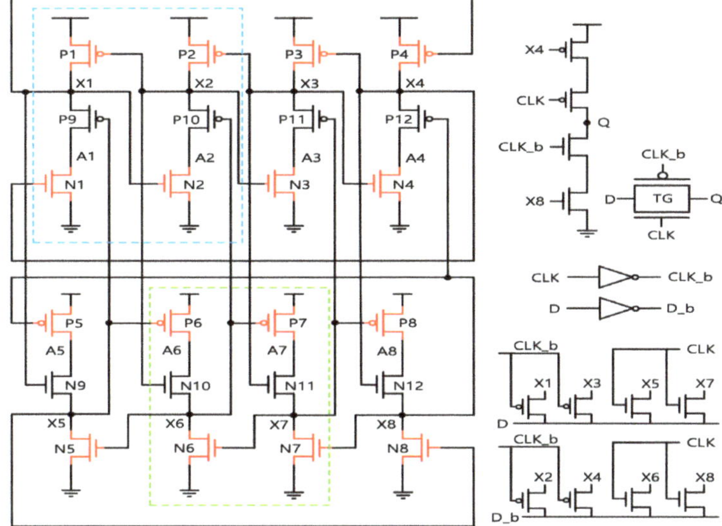

Figure 2. The proposed fully polarity-aware DNU-resilient latch (FPADRL) (when X1 = X3 = X5 = X7 = 1, X2 = X4 = X6 = X8 = 0, and D = Q = 1).

Because of the stacked MOSFETs in the pull-down (up) paths, all the data nodes have the body effect, i.e., V_{th} drop. To reduce the leakage power by decreasing the overdrive voltage, the proposed latch uses multiple V_{th} transistors. The transistors shown in red in Figure 2 use high V_{th}, and others use low V_{th}. Additionally, the leakage power can be minimized by connecting one strong signal node to the off-transistor and one weak state node to the on-transistor of the output inverter. The high V_{th} transistors shrink the activated feedback loop time, resulting in a shorter write time compared to the DICE-based DNU latches.

2.2. Circuit Operation

The operation of the proposed latch is outlined when the input D is logic "1" (i.e., X1 = X3 = X5 = X7 = 1 and X2 = X4 = X6 = X8 = 0). Considering the error polarity, the access transistors of the PSLP are pMOSFETs, and the access transistors of the NSLP are nMOSFETs. When CLK = 1, the latch is in the transparent mode, and these access transistors drive the internal nodes. During this mode, the output Q is only driven by D through the transmission gate. The driven nodes turn on the nMOSFETs (N2, N4, N6, N8, N9, and N11) and the pMOSFETs (P1, P3, P5, P7, P10, and P12) simultaneously. As a result, nodes X1 and X3 are strong "1" state, and the nodes X2 and X4 are weak "0." In contrast, nodes X5 and X7 are weak "1" state, and nodes X6 and X8 are strong "0." In the CLK = 0 phase, the latch is in the hold mode, the access transistors and the transmission gate are opened, and the clock-gated inverter drives the output Q using the logic values of the internal nodes. When D = 1, the data nodes X2, X4, X5, and X7 and the floating nodes A1, A3, A6, and A8 are SEU-sensitive. Likewise, when D = 0, the latch operates complementary.

2.3. SEU-Resilience Analysis

As explained in Section 2.1, the proposed latch reduces the SEU-sensitive internal nodes as fully as possible, from sixteen to eight. Moreover, the sensitive nodes have not both-way transitions but one-way error polarity with 0-to-1 or 1-to-0 transitions. Therefore, the proposed latch has 8 SNU nodes and 28 DNU node pairs. As shown in Table 1, the proposed latch has ten different upset cases. Each case includes example node(s), SEU polarity and recovery mechanism; "↑" ("↓") denotes 0 (1)-to-1 (0) SEU transition, "←" and "→" denote that SEU propagates to that node, "↚" and "↛" denote that SEU does not propagate to that node. Table 1 shows only the cases when D = 1. The SEU recovery of the complementary input, D = 0, has the same recovery mechanism since the circuit topology is symmetric.

When an ionizing particle strikes at the floating node, like Case 2 in Table 1, the upset does not affect data nodes since the floating node is not connected to any inputs. Considering Case 2 mechanism, Case 4 and Case 5, DNU between data node and adjacent (a) or remote (r) floating node, can be simplified to Case 1. In Case 6 and Case 7, the upset occurs on the node that controls the off-stacked transistors, so the error propagates to the adjacent data node and temporarily creates the short (s) path. However, the proposed latch can recover in these cases. For example, in Case 7, the error-propagated node X3's voltage does not turn off N11, even though node X3 is short. Therefore, only the recover-charging current through P7 and N11 flows to node X7, causing node X3's race to nullify by turning off P11 gradually. The recovery of node X3 results in the recovery of node X2. Like Case 8, deposited charge on a floating node A6 can be shared with a data node X6 by turning on N10. In this case, the node X5 voltage can turn on P6 and P9. However, since those weak inversion transistor currents are smaller than the strong driven currents from the neighboring transistors, the logic value of node X1 does not change, and node X6 can be recovered. The proposed latch can also recover cases when the SNU or DNU occurs at the output node Q since the proposed latch guarantees the SNU resiliency of all internal nodes. Consequently, the proposed latch is resilient to all SNU and DNU cases. The detailed transistor sizing scheme for SEU-resilience will be explained in Section 3.1.

2.4. Radiation-Aware Layout

Figure 3 shows the proposed layout considering layout-level issues such as incidence angle, charge sharing (CS), and parasitic bipolar effect (PBE) [2]. Sensitive nodes are colored in orange when D = 1. The proposed layout utilizes double height cell design [12]: N-well at the top and bottom, P-well in the middle. The middle P-well acts as a canceling area between two N-wells. Based on this principle, the chance of multiple-node-upsets in different wells (severe cases such as Case 6–8) can be lowered. Moreover, our node placement can mitigate CS and PBE in sensitive data nodes. We separated NSLP's two sensitive data nodes as far as possible in P-well. Two sensitive data nodes in PSLP are in the different N-well. To reduce the effective amount of injected charge, we placed on- and off-transistors repeatedly. Off-transistors with Ax nodes act as a strong collector, and on-transistors with Ax nodes act as a weak collector when the upset occurs on the data node. Therefore, the data node's upset threshold increases [13]. However, we need to reconfigure the layout in more advanced technology nodes because of the polysilicon bends.

Table 1. SEU-Resilience Mechanism (D = 1).

Case: Node(s)	Example	Polarity	Recovery Mechanism
1: data	X2	↑	X1(P1-off) ↚ X2 → A3(N3-on) ↛ X3(P11-off) ⇒ X2↓ (P2-off, P10, N2-on)
2: floating	A1	X	A1(∄ connected input) ↛ X1(P9-off)
3: data pair (same part)	X2, X4	↑, ↑	X1(P1-off) ↚ X2 → A3(N3-on) ↛ X3(P11-off) X3(P3-off) ↚ X4 → A1(N1-on) ↛ X1(P9-off) ⇒ X2↓ (P2-off, P10 and N2-on) and X4↓ (P4-off, P12 and N4-on)
4: data, (a)floating (same part)	X2, A1	↑, X	A1(∄ connected input) ↛ X1(P9-off) ⇒ Case 4 ≈ Case 1
5: data, (r)floating (same part)	X2, A3	↑, X	A3(∄ connected input) ↛ X3(P11-off) ⇒ Case 5 ≈ Case 1
6: data pair propagation in NSLP (different part)	X2, X5	↑, ↓	X1(P1-off) ↚ X2 → A3(N3-on) ↛ X3(P11-off) X8(N8-off) ↚ X5 → A6(P6-on) → X6↑ (s, N10-on) ↛ A7(P7-off) ⇒ X2↓ (P2-off, N2 and P10-on) (∵ $V_{X6\uparrow} < V_{DD} - V_{thp_L}$) ⇒ X6↓ (P6 and N10-off, N6-on) ⇒ X5↑ (N5-off, P5 and N9-on)
7: data pair propagation in PSLP (different part)	X2, X7	↑, ↓	X1(P1-off) ↚ X2 → A3(N3-on) → X3↓ (s, P11-on) ↛ A4(N4-off) X6(N6-off) ↚ X7 → A8(P8-on) ↛ X8(N12-off) ⇒ X7↑ (N7-off, P7 and N11-on) (∵ $V_{X3\downarrow} > V_{thn_L}$) ⇒ X3↑ (P11 and N3-off, P3-on) ⇒ X2↓ (P2-off, P10 and N2-on)
8: data, (a)floating (different part)	X2, A6	↑, X	X1(P1-off) ↚ X2 → A3(N3-on) ↛ X3(P11-off) X5↓ (s, N5-on) ← X6 ↑ (by A6, N10-on) ↛ A7(P7-off) ⇒ X6↓ (P6-on, N10 and N6-on) (∵ P6 weak inversion) ⇒ X2↓ (P2-off, P10 and N2-on) and X5↑ (N5-off, P5 and N9-on)
9: data, (r)floating (different part)	X2, A8	↑, X	A8(∄ connected input) ↛ X8(N12-off) ⇒ Case 9 ≈ Case 1
10: floating pair	A1, A3	X, X	A1(∄ connected input) ↛ X1(P9-off) A3(∄ connected input) ↛ X3(P11-off)

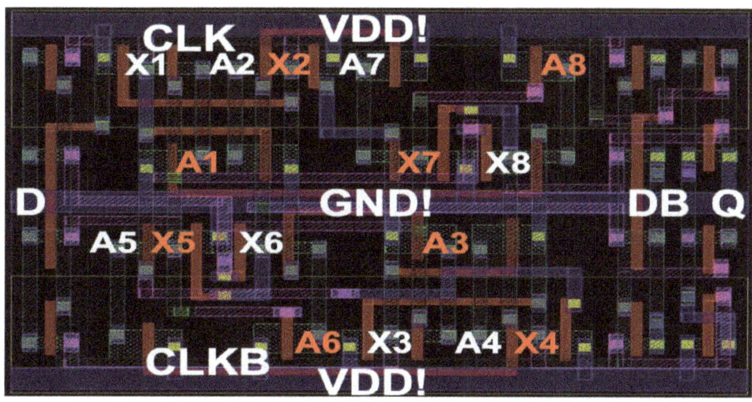

Figure 3. The proposed double-height radiation-aware layout of FPADRL (when X1 = X3 = X5 = X7 = 1, X2 = X4 = X6 = X8 = 0, and D = Q = 1).

3. Evaluation Results

3.1. Radiation Simulation Results

The proposed latch was designed using the 45-nm NCSU CMOS technology. Due to its unique recovery process, the DICE-based latch provides a large margin of transistor sizes while keeping the SEU resilience. Therefore, we can aggressively optimize the sizes with minimum PDP. However, this approach may result in many different transistor sizes not suitable for the state-of-the-art layout. Considering both PDP and layout, we decided on smaller groups of uniform size for transistor sizes. Table 2 shows the transistor sizes, in which the aspect ratio of 1 is W/L = 90 nm/50 nm. The low (high) V_{th} of n(p)MOSFET are 0.322 V (-0.302 V) and 0.608 V (-0.505 V), respectively.

To validate the proposed latch's SEU-resilience, the dual-double exponential current source [14] model was used instead of the conventional double exponential upset model because it provides a current shape similar to the actual SEU-current. The simulations were performed using Smartspice from Silvaco. Figure 4 shows the radiation simulation waveforms for all the cases in Table 1. These waveforms show that the output Q is always error-free, and all the internal nodes are recovered in all SNU and DNU cases. Therefore, it is clearly demonstrated that the proposed latch is resilient to both SNU and DNU.

Table 2. Transistor Characteristics.

Transistor	Aspect Ratio	V_{th}
P1–P4	2.5	High
P9–P12	1	Low
N1–N4	2	High
P5–P8	2	High
N9–N12	1	Low
N5–N8	2.5	High

3.2. Performance Comparison and Evaluation

Table 3 shows the comparison results at the TTTT (1.1 V/25 °C/TT). Using the same technology with the proposed latch, the referred designs were re-implemented with the same size ratio the corresponding manuscripts provided. The simulation was conducted under a clock frequency of 100 MHz. The clock frequency was set enough for checking normal operation and recovery operation simultaneously.

Table 3. Performance Comparison results of DNU hardened circuits.

		[5]	[6]	[7]	[8]	[9]	[10]	FPADRL
# of Transistors		66	48	42	38	28	36	**42**
# of Nodes		21	24	10	10	6	12	**16**
# of Sensitive nodes		21	24	10	10	6	9	**8**
Area (μm^2)		10.48	N/A	11.63	8.80	N/A	10.98	**9.69**
t_{dq} (ps)		5.90	6.57	22.79	33.86	2.70	2.70	**2.66**
t_{setup} (ps)		52.65	16.21	14.02	21.28	30.62	62.78	**40.73**
Power (μW)	Opaque	0.71	0.47	0.44	0.33	0.19	4.14	**0.43**
	DAR 100%	2.90	1.84	2.26	3.21	1.29	7.95	**1.55**

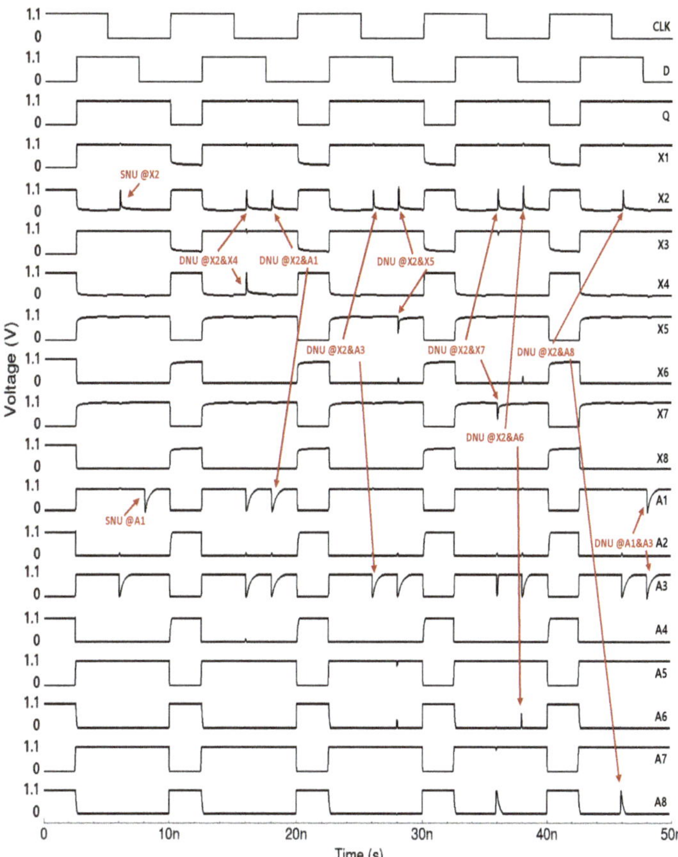

Figure 4. Radiation simulation waveforms of all the SNU and DNU cases.

By stacking the transistors with regular latch architecture, the proposed latch fully reduces the SEU-sensitive nodes by half from sixteen to eight. Moreover, among the eight sensitive nodes, four floating nodes are less sensitive to SEU, and the other four nodes are one-way sensitive, not both ways. The area of the proposed layout is compared based on the scaled-down results of [10]. Even though the proposed layout is radiation-aware and others are not, its regular latch architecture makes 12% less area compared with the state-of-the-art polarity-aware latch [10]. Although the number of transistors in the FPADRL is not the lowest, its layout area can be easily optimized because of its lower design complexity than regular latch architecture. In terms of delay, t_{dq} and t_{setup} are evaluated. Power consumption profiles are divided into static power during opaque mode and dynamic power in DAR of 100%.

The MCE-based DNU latches [5,6] require extra transistors to achieve upset resilience. Fundamentally, the MCE relies on high impedance for radiation hardening. However, because of this property, the MCE-extended design usually needs a plethora of transistors for resilience. When the upset occurs, the MCE blocks not only error propagation to adjacent nodes but also the feedback path for recovery. Therefore, [5,6] consist of 66 and 48 transistors, which are the top two largest number of transistors. The more transistors used, the higher the power consumption. In terms of delay, [5,6] have a moderate speed compared to the DICE-extended design [7,8] due to directly driven output Q through the transmission gates. However, the input D should drive the nodes with large capacitance, so its delay is longer than our proposed design. As a result, [5,6] have up to 60% delay and 47% power overhead compared with the proposed FPADRL.

The DICE-extended designs [7,8] have a considerable delay and power consumption. Because they inherit DICE, the designs should activate all DICE feedback loops when the data are switching. It indicates that input D should have the large driving capability to write new data, resulting in performance penalties. Moreover, the input D does not drive the output Q directly, so the t_{dq} delay is more considerable than other reference designs. As a result, the proposed FPADRL achieves a reduced delay and power (in DAR 100%) up to 92% and 52%, respectively, compared with [7,8].

The ST-extended design [9] has similar delay and power compared to the proposed FPADRL. The design uses the least number of transistors among the reference designs and our proposed design. However, the design is susceptible to the ratio issue. Hence, its radiation-hardening capability is weak, as presented in Section 3.3.

Lastly, the conventional polarity-aware latch [10] consumes considerable power due to its irregular cross-coupled latch structures with only single V_{th} transistors. By using regular latch structures and high-threshold voltage transistors to make shorter activated feedback time and to minimize the leakage power, the proposed FPADRL achieves 80% power (in DAR 100%) reduction compared with [10].

Regarding setup time, the proposed latch pales in comparison to [6–9]. Like the DICE-extended designs [7,8], FPADRL activates all feedback loops when writing new data. However, the setup performance is degraded due to the high V_{th} device and the stacked MOSFETs. The MCE-extended design [5] and the polarity-aware latch [10] require a longer setup time than the proposed latch because of the circuit topology for DNU-resilience and irregular latch structure.

For a more detailed evaluation of the power consumption, the comparison was made in the range of the DAR 0% to 100% at the TTTT. Figure 5 shows the proposed latch outperforms power against the MCE-based DNU latches [5,6] and the DICE-based DNU latches [7,8]. Because of the V_{th} drop issue, the recent polarity-aware latch [10] shows huge power consumption than all other circuits. Although the ST-based DNU latch [9] is shown to have superiority in the power over the proposed latch, the ratio issue of the ST's recovery yields inferior radiation-hardening capability, as shown in Section 3.3.

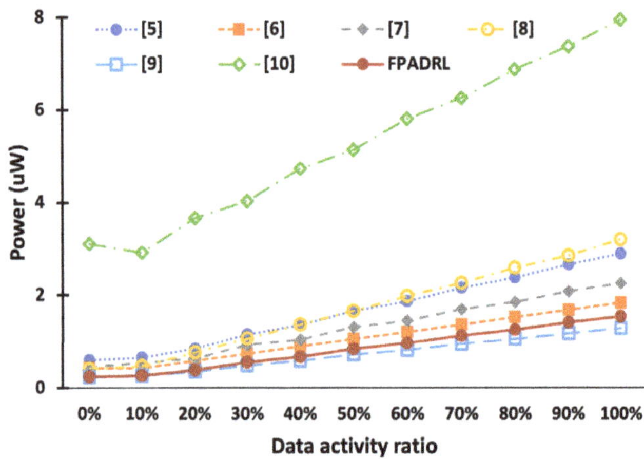

Figure 5. Power consumption under different data activity ratio (DAR).

A simulation with the process, voltage, and temperature (PVT) corner analysis was performed according to the commercial standard. The following three extreme conditions were set: (1) SSSS: 0.99 V/125 °C/SS, (2) TTTT: 1.1 V/25 °C/TT, (3) FFFF: 1.21 V/−40 °C/FF.

Figure 6 shows the performance comparison with the PVT corner analysis. From the reasons described in the previous paragraph, the DICE-based DNU latches [7,8] have the most extensive t_{dq} variation, and the MCE-based DNU latches [5,6] follow next. Although the conventional polarity-aware latch [10] has a shorter delay and smaller variation, [10] consumes more power. The ST-based latch [9] has the lowest delay and power due to its simple architecture. However, it suffers from the ratio issue for the radiation hardening. FPADRL has longer setup time in three conditions than [6–9]. Note that the setup time of the proposed latch increases significantly in the SSSS corner due to the high V_{th} device like [9].

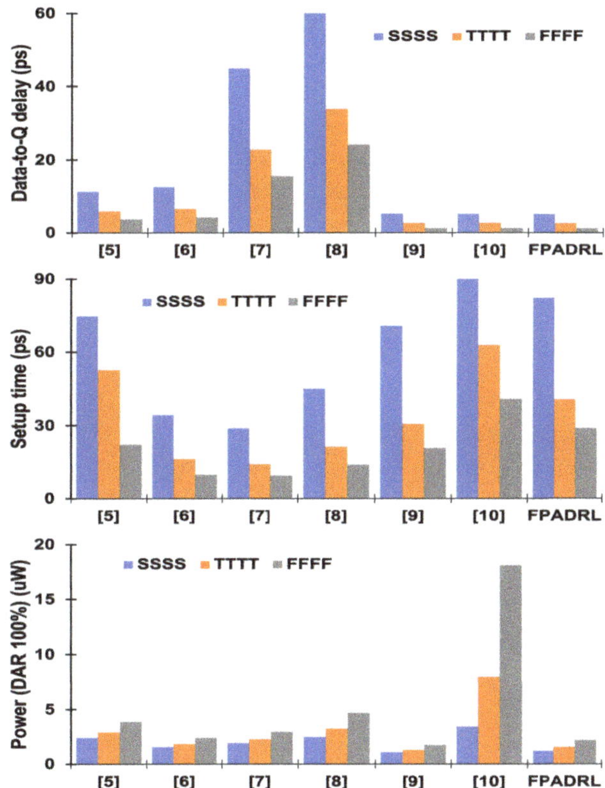

Figure 6. Performance Comparison in PVT corner analysis.

3.3. Radiation-Hardening Capability Comparison

The robustness of the proposed latch against SEU was simulated and compared with those of other latches. In order to analyze the effect of robustness according to the process corner, the simulation was conducted for five corners at 1.1 V/25 °C: SS, SF, TT, FS, FF. For the comparison, different amounts of charge according to the inverter size (INVX) were injected into key node pairs of each design. Table 4 shows the peak and plateau currents of the DDECS and the total amount of charge. Given that the SEU width depends on the LET radiation [15,16], the SEU width of the INV1, INV2, and INV4 were set to be 50 ps, 100 ps, and 200 ps, respectively. The INV1's nMOSFET size is W/L = 90 nm/50 nm and pMOSFET size is W/L = 180 nm/50 nm. The INV2 is the double size of the INV1, and the INV4 is the quadruple size of the INV1.

Using the various injection parameter settings in Table 4, the current was injected to key node pairs. The failure probability is calculated by the number of cases in which DNU are not recovered among all cases. Figure 7 shows the failure probability among the DNU-resilient latches with different corners. The ST-based [9] shows the weakest radiation-hardening capability. In contrast, the proposed FPADRL and [8] can recover every case in all corners.

Table 4. Injected charge configuration.

Cell Name	Load	Polarity	I_{Peak-p} (μA)	I_{Peak-h} (μA)	Q_{total} (fC)	SEU Width (ps)
INV1	INV1	$\uparrow$	46	120	8.41	50
		$\downarrow$	41	162	10.91	
INV2	INV2	$\uparrow$	127	121	16.22	100
		$\downarrow$	126	164	21.01	
INV4	INV4	$\uparrow$	152	122	29.06	200
		$\downarrow$	159	164	38.11	

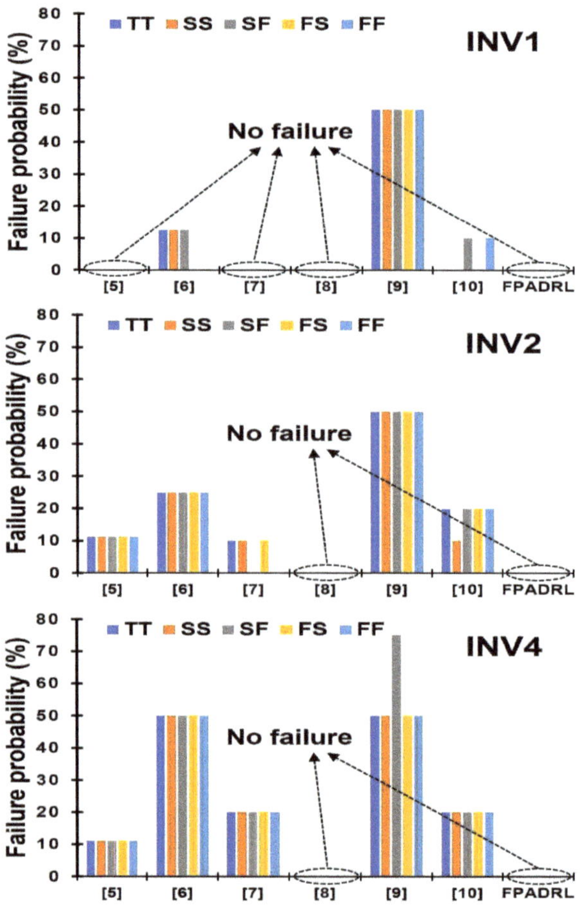

Figure 7. Failure probability comparison in different process corner.

Table 5 shows the recovery power at the TTTT. The recovery power is defined as the power difference between upsets and normal operation (no upsets). Although the

MCE-based DNU latches [5,6] consume less recovery power than FPADRL, FPADRL overwhelms the radiation-hardening capability than [5,6] as shown in Figure 7. Unlike the DICE-based DNU latches [7,8] which are suffered from the short path when the upset occurs, the proposed latch does not create the short path in most cases due to the stacked MOSFETs. Thus, FPADRL consumes less power for recovery than [8] which has the same radiation-hardening capability.

Table 5. Radiation-hardening Comparison of DNU-Resilient Latches.

		[5]	[6]	[7]	[8]	[9]	[10]	**FPADRL**
INV1	Recovery Power (μW)	5.67	8.16	14.98	13.31	8.92	14.96	**14.29**
INV2	Recovery Power (μW)	17.55	19.07	31.05	30.65	20.29	19.46	**24.35**
INV4	Recovery Power (μW)	54.89	38.58	53.99	55.93	45.64	34.28	**41.01**

To summarize, FPADRL is the best candidate among state-of-the-art DNU-resilient latches in a comprehensive view comparison. Although the FPADRL does not show the best performance, it has the strongest radiation-hardening capability with low recovery power.

4. Conclusions

A fully polarity-aware DNU-resilient latch with high performance and low power was proposed. The proposed latch is self-recoverable for all SNU and DNU cases. Its performance and radiation-hardening capability comparison against other latches indicate that the proposed FPADRL is performance-effective and highly robust with less overhead. Consequently, the proposed FPADRL exhibits certain advantages compared with state-of-the-art DNU-resilient latches.

Author Contributions: Conceptualization, J.-J.P., Y.-M.K. and J.K.; methodology, J.-J.P. and Y.-M.K.; software, J.-J.P. and Y.-M.K.; validation, J.-J.P., Y.-M.K. and G.-H.K.; formal analysis, J.-J.P., Y.-M.K. and J.K.; investigation, J.-J.P., Y.-M.K., G.-H.K., I.-J.C. and J.K.; resources, J.-J.P., Y.-M.K. and J.K.; data curation, J.-J.P. and J.K.; writing—original draft preparation, J.-J.P.; writing—review and editing, J.-J.P., Y.-M.K., G.-H.K., I.-J.C. and J.K.; visualization, J.-J.P., Y.-M.K. and G.-H.K.; supervision, I.-J.C. and J.K.; project administration, I.-J.C. and J.K.; funding acquisition, J.K. All authors have read and agreed to the published version of the manuscript.

Funding: This research was funded by National Research Foundation of Korea, grant number: 2019R1F1A1054130 and Institute for Information and Communications Technology Promotion, grant number: 2021001764.

Institutional Review Board Statement: Not applicable.

Informed Consent Statement: Not applicable.

Data Availability Statement: The data presented in this study are available on request from the corresponding author. The data are not publicly available due to licensing agreements.

Conflicts of Interest: The authors declare no conflict of interest.

References

1. Pudi N S, A.K.; Baghini, M.S. Robust Soft Error Tolerant CMOS Latch Configurations. *IEEE Trans. Comput.* **2016**, *65*, 2820–2834. [CrossRef]
2. Black, J.D.; Dodd, P.E.; Warren, K.M. Physics of multiple-node charge collection and impacts on single-event characterization and soft error rate prediction . *IEEE Trans. Nucl. Sci.* **2013**, *60*, 1836–1851. [CrossRef]
3. Calin, T.; Nicolaidis, M.; Velazco, R. Upset hardened memory design for submicron CMOS technology. *IEEE Trans. Nucl. Sci.* **1996**, *43*, 2874–2878. [CrossRef]
4. Loveless, T.; Jagannathan, S.; Reece, T.; Chetia, J.; Bhuva, B.; McCurdy, M.; Massengill, L.; Wen, S.J.; Wong, R.; Rennie, D. Neutron-and proton-induced single event upsets for D-and DICE-flip/flop designs at a 40 nm technology node. *IEEE Trans. Nucl. Sci.* **2011**, *58*, 1008–1014. [CrossRef]

5. Yan, A.; Huang, Z.; Yi, M.; Xu, X.; Ouyang, Y.; Liang, H. Double-node-upset-resilient latch design for nanoscale CMOS technology. *IEEE Trans. Very Large Scale Integr. (VLSI) Syst.* **2017**, *25*, 1978–1982. [CrossRef]

6. Yan, A.; Yang, K.; Huang, Z.; Zhang, J.; Cui, J.; Fang, X.; Yi, M.; Wen, X. A Double-Node-Upset Self-Recoverable Latch Design for High Performance and Low Power Application. *IEEE Trans. Circuits Syst. II Express Briefs* **2019**, *66*, 287–291. [CrossRef]

7. Eftaxiopoulos, N.; Axelos, N.; Zervakis, G.; Tsoumanis, K.; Pekmestzi, K. Delta DICE: A double node upset resilient latch. In Proceedings of the 2015 IEEE 58th International Midwest Symposium on Circuits and Systems (MWSCAS), Fort Collins, CO, USA, 2–5 August 2015; IEEE: Piscataway, NJ, USA, 2015; pp. 1–4.

8. Eftaxiopoulos, N.; Axelos, N.; Pekmestzi, K. DONUT: A double node upset tolerant latch. In Proceedings of the 2015 IEEE Computer Society Annual Symposium on VLSI, Montpellier, France, 8–10 July 2015; IEEE: Piscataway, NJ, USA, 2015; pp. 509–514.

9. Li, Y.; Cheng, X.; Tan, C.; Han, J.; Zhao, Y.; Wang, L.; Li, T.; Tahoori, M.B.; Zeng, X. A Robust Hardened Latch Featuring Tolerance to Double-Node-Upset in 28nm CMOS for Spaceborne Application. *IEEE Trans. Circuits Syst. II Express Briefs* **2020**, *67*, 1619–1623. [CrossRef]

10. Guo, J.; Liu, S.; Zhu, L.; Lombardi, F. Design and evaluation of low-complexity radiation hardened CMOS latch for double-node upset tolerance. *IEEE Trans. Circuits Syst. I Regul. Pap.* **2020**, *67*, 1925–1935. [CrossRef]

11. Kelin, L.H.H.; Klas, L.; Mounaim, B.; Prasanthi, R.; Linscott, I.R.; Inan, U.S.; Subhasish, M. LEAP: Layout design through error-aware transistor positioning for soft-error resilient sequential cell design. In Proceedings of the 2010 IEEE International Reliability Physics Symposium, Anaheim, CA, USA, 2–6 May 2010; IEEE: Piscataway, NJ, USA, 2010; pp. 203–212.

12. Uemura, T.; Tosaka, Y.; Matsuyama, H.; Shono, K.; Uchibori, C.J.; Takahisa, K.; Fukuda, M.; Hatanaka, K. SEILA: Soft error immune latch for mitigating multi-node-SEU and local-clock-SET. In Proceedings of the 2010 IEEE International Reliability Physics Symposium, Anaheim, CA, USA, 2–6 May 2010; IEEE: Piscataway, NJ, USA, 2010; pp. 218–223.

13. Kawakami, Y.; Hane, M.; Nakamura, H.; Yamada, T.; Kumagai, K. Investigation of soft error rate including multi-bit upsets in advanced SRAM using neutron irradiation test and 3D mixed-mode device simulation. In Proceedings of the IEDM Technical Digest. IEEE International Electron Devices Meeting, San Francisco, CA, USA, 13–15 December 2004; IEEE: Piscataway, NJ, USA, 2004; pp. 945–948.

14. Black, D.A.; Robinson, W.H.; Wilcox, I.Z.; Limbrick, D.B.; Black, J.D. Modeling of single event transients with dual double-exponential current sources: Implications for logic cell characterization. *IEEE Trans. Nucl. Sci.* **2015**, *62*, 1540–1549. [CrossRef]

15. Gadlage, M.; Schrimpf, R.; Benedetto, J.; Eaton, P.; Mavis, D.; Sibley, M.; Avery, K.; Turflinger, T. Single event transient pulse widths in digital microcircuits. *IEEE Trans. Nucl. Sci.* **2004**, *51*, 3285–3290. [CrossRef]

16. Cohn, L. Single-event effects in advanced digital and analog microelectronics. In Proceedings of the Microelectronics Reliability and Qualification Workshop, Manhattan Beach, CA, USA, 4–5 December 2007 .

Article

TID Sensitivity Assessment of Quadrature LC-Tank VCOs Implemented in 65-nm CMOS Technology

Arijit Karmakar *, Valentijn De Smedt and Paul Leroux

Department of Electrical Engineering (ESAT)-ADVISE, KU Leuven, 2440 Geel, Belgium;
valentijn.desmedt@kuleuven.be (V.D.S.); paul.leroux@kuleuven.be (P.L.)
* Correspondence: arijit.karmakar@kuleuven.be

Abstract: This article presents a comprehensive assessment of the ionizing radiation induced effects on the performance of quadrature phase LC-tank based voltage-controlled-oscillators (VCOs). Two different quadrature VCOs (QVCOs) that are capable of generating frequencies in the range of 2.5 GHz to 2.9 GHz are implemented in a commercial 65 nm bulk CMOS technology to target for harsh radiation environments like space applications and high-energy physics (HEP) experiments. Each of the QVCOs consumes 13 mW power from a 1.2 V supply. The architectures are based on the popular implementation of two different types of QVCOs: parallel-coupled QVCO (PQVCO) and super-harmonic coupled QVCO (SQVCO). The various performance metrics (oscillation frequency, quadrature phase, phase noise, frequency tuning range, and power consumption) of the two different QVCOs are evaluated with respect to a Total ionizing Dose (TID) up to a level of approximately 100 Mrad (SiO$_2$) through X-ray irradiation. During irradiation, the electrical characterization of the samples of the prototype are performed under biased condition at room temperature. Before irradiation, the QVCOs (PQVCO and SQVCO) achieve phase noise equal to -115 dBc/Hz and -119 dBc/Hz at 1 MHz offset, resulting in figure-of-merit (FoM) of -172.2 dBc/Hz and -176.4 dBc/Hz respectively. The test-setup of the TID experiment is discussed and the results obtained are statistically analyzed in this article to perform a comparative study of the performance of the two different QVCOs and evaluate the effectiveness of the radiation hardened by design techniques (RHBDs) employed in the implementations. Post-irradiation, the overall variations of the frequencies of the oscillators are less than 1% and the change in tuning range (TR) is less than 5% as observed from the tested samples.

Keywords: quadrature; super-harmonic; LC-tank; Q-phase; VCO; QVCO; radiation; TID; SEE; X-ray; high energy physics; radiation hardened by design

Citation: Karmakar, A.; De Smedt, V.; Leroux, P. TID Sensitivity Assessment of Quadrature LC-Tank VCOs Implemented in 65-nm CMOS Technology. *Electronics* **2022**, *11*, 1399. https://doi.org/10.3390/electronics11091399

Academic Editor: Anna Richelli

Received: 25 March 2022
Accepted: 26 April 2022
Published: 27 April 2022

Publisher's Note: MDPI stays neutral with regard to jurisdictional claims in published maps and institutional affiliations.

1. Introduction

Various high frequency designs predominantly in the area of radio-frequency and millimeter-wave based applications [1] require quadrature phase shifted signals to enable wireless as well as wired communication systems. Wireless communication systems with integrated transceivers require quadrature signals for up and down conversion and eventually require accurate quadrature phase for effective image-rejection in the base-band [2–4]. In the case of wired communication, QVCOs play a key role in multi-phase clock generation and in particular assist in the implementation of the half-rate clock and data recovery (CDR) circuits [5,6]. Additionally, QVCOs have also been explored to be integrated in phased-array transceivers [7,8].

Numerous integrated design techniques [9,10] for quadrature phase generation are reported in the literature to date and can be broadly identified as using (1) active R-C oscillators [11], (2) relaxation oscillators [12], (3) ring oscillators [13,14], (4) poly-phase filters [15,16], (5) frequency divide-by-2 circuits [17], (6) cross-coupled QVCOs, and (7) super-harmonic coupled QVCOs. Amid these, LC-tank oscillator-based options have prevailed as the design choice considering its superior performance in terms of phase noise and spectral purity within the given power budget. However, these are achieved with significant area penalty due to on-chip inductors. Other non-LC-tank based options are explored, despite the fact that, in some low-frequency applications they are limited by the area constraints and they do not have particularly stringent phase noise requirements. LC-tank based QVCOs typically contain two identical LC oscillators with the outputs cross coupled with each other. The quadrature outputs can be coupled through active devices [9,18–20] or passive devices [21,22]. The former method improves phase accuracy but at the cost of increased phase noise, whereas the latter showcases improved phase noise contribution but trades off phase accuracy due to limited coupling strength. Conventionally, the coupling mechanism in the cross-coupled QVCOs follow either parallel [9,20] or series [18] coupling schemes using the fundamental frequency component. The existing trade-off between phase noise and phase accuracy among these architectures can be eliminated by achieving quadrature phase locking using super-harmonic coupling, i.e., second harmonic injection at 180° out-of-phase at the oscillators' common-mode nodes [10].

The use of QVCOs is extensive in various communication systems and half-rate CDR circuits. In modern days, these oscillators have also found indispensable use in high-speed communication in space (satellite communication, on-board spacefibre network) applications and data transmission during HEP experiments. Considering the harsh radiation environment these oscillators are subjected to, the oscillators required to sustain up to a TID level of several Mrad for space applications but several hundreds of Mrad for HEP experiment. In CMOS technology, the effects of radiation on metal–oxide–semiconductor field-effect transistor (MOSFET) devices are multifold. While exposed to continuous radiation, the performance of the devices is degraded with the change in threshold voltage, drain current, intrinsic gain and noise levels [23]. The radiation induced effects in presence of mismatch and non-linearity degenerate the system's performance, and in many of the cases, striking radiation particles lead to functional failures and transients at the output. Radiation strike in the gate-oxide region of MOSFET devices generates pairs of electron–hole and its numbers are proportional to the energy deposited during the interaction and inversely proportional to the squared value of its gate-oxide thickness (t_{ox}). The oxide-trap charges result in an increase in threshold voltage for p-channel MOS (PMOS) devices and a decrease for n-channel MOS (NMOS) devices. This effect in NMOS devices is counter-acted by interface trap charges, but aggravated in the case of PMOS devices. With the scaling of CMOS technologies, the reduction in gate-oxide is likely to lessen the TID induced problems, but on the contrary has increased SEE induced transients due to the lower critical charge. However, in lower feature size (sub-90 nm) dense technologies, the accumulated charges in shallow trench isolation (STI) regions still affect the device performance [24–26] over the course of time.

As reported in the literature, the performance of LC-tank based single VCOs operating under ionizing radiation are extensively studied in [27,28]. The vulnerability of VCO architectures with respect to TID effects for space applications as well as in HEP experiments are explored in [29], whereas the effects of radiation-induced single-event-effects (SEEs) in VCOs are studied and a few possible remedies are suggested with some design improvements in [28,30,31]. In comparison, the knowledge about the effects of ionizing radiation on the performance of QVCOs are limited due to scarcity in the number of studies [32,33] done on QVCOs under radiation exposure. The SEE induced effects are studied in [32] on QVCOs implemented in 65 nm bulk CMOS technology. The TID induced effects on a parallel-coupled QVCO implemented in 32 nm silicon-on-insulator (SOI) CMOS technology is explored in [33]. The study reports on the degradation of several performance parameters (oscillation frequency, phase noise and power consumption), but gives no insight on the radiation induced effects on quadrature phase accuracy. This article focuses on giving more insights on the effects of radiation on QVCOs with a particular interest in TID induced variations. To the authors' knowledge, for the first time to date, an experimental study and comparison of TID induced performance variation is done between two different topologies of LC-tank based QVCOs implemented in 65 nm bulk CMOS technology.

The rest of the article is arranged as follows: Section 2 discusses the implementation of the QVCOs in detail. The test-setup for X-ray irradiation is elaborated in Section 3 together with a detailed description of the results of the experiment and a comparison of performance studies of the two different QVCOs. Section 4 concludes the article with the summary of the TID sensitivity study on the two QVCOs.

2. Implementation

For the TID experiment, two different topologies of QVCOs are implemented for quadrature phase generation: parallel coupled QVCO (PQVCO) and super-harmonic coupled QVCO (SQVCO).

A schematic diagram of the PQVCO as proposed in [34] with two identical core structures is shown in Figure 1. Each oscillator core is implemented with complementary cross-coupled PMOS-NMOS transistor pairs (M_P, M_N) with an LC-tank. The output of each oscillator is coupled to the other's output nodes parallelly using NMOS transistor (M_R) pairs. Based on theory explained in [35], if the two oscillator cores are symmetrical and matched in terms of parasitics, the four outputs V_{A+}, V_{A-}, V_{B+}, V_{B-} produce quadrature phase shifted differential In-phase and Q-phase (I and Q) output signals. The tunable capacitors in the LC-tank circuit are realised using n-well MOS varactors (C_{var}) and the inductor is implemented in top-metal to maximize the inductor quality factor. As shown in Figure 2a, the n-well MOS varactors are ac-coupled to the oscillator's output nodes using capacitors ($\sim5 \times C_{var}$) and the gate-voltage is varied using V_F connected through a resistor. This arrangement helps to reduce the sensitivity towards single event transients (SETs) [36]. Here, any charge collected between the n-well and p-substrate interface finds a low impedance path to ground through the metal wires. However, this improvement pays a penalty of reduced frequency tuning range (TR). Also wider devices are chosen for core PMOS-NMOS pairs (M_P, M_N) to minimize radiation induced performance variation [37], in particular to reduce parasitic turn-on of the lateral devices due to STI-trapped charges. However, the use of wider devices results in increased power consumption. The oscillators are designed to operate with full swing with large bias current (~10 mA) at the edge of the current limited region to achieve optimum phase noise performance within the power budget. The bias current is generated by using typical current-mirror circuits formed with NMOS pairs. The bottom-biasing using NMOS is chosen instead of top-biased PMOS as the radiation induced variation is less in NMOS devices compared to PMOS devices in 65 nm CMOS technology [37].

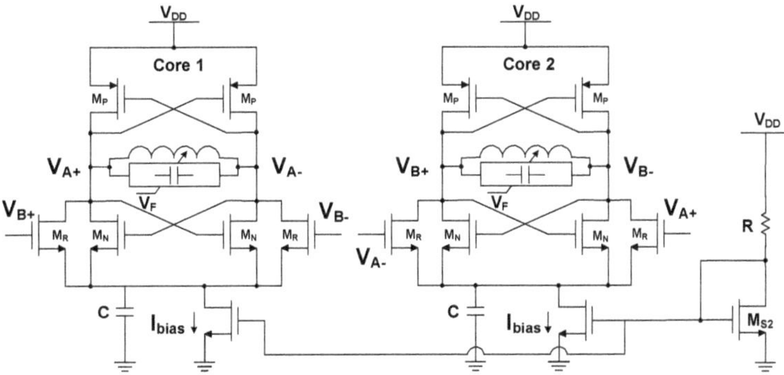

Figure 1. A schematic representation of the parallel coupled quadrature LC-tank VCO (PQVCO).

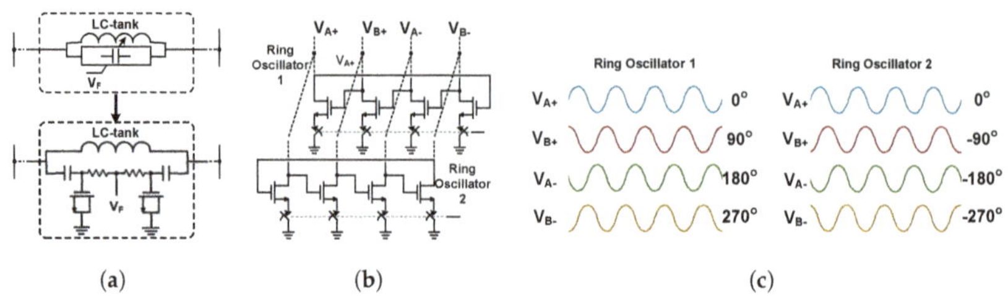

Figure 2. (a) Tunable LC-tank circuit with ac-coupled NMOS varactor, (b) ring oscillators for phase lead/lag selection, and (c) oscillator generated output waveforms for different ring oscillator configurations.

A simplified block diagram of the PQVCO cross-coupled with each other through transconductance ($g_{m,c}$) formed by the coupling NMOS transistors (M_R) is shown in Figure 3. The LC-tank circuits (R_p, C, L) form the tank band-pass filter at the resonance frequency and complementary PMOS-NMOS pairs (M_P, M_N) create the $-1/g_m$ cell to compensate for tank losses. Quadrature phase accuracy can be improved with stronger coupling strength $g_{m,c}$, but it leads to increased phase noise around the oscillation frequency. Therefore, the sizes of M_R are chosen smaller (1/4 times) with respect to the core MOS devices (M_N) to minimize the phase noise contribution while maintaining the quadrature phase error within acceptable limits ($<1°$). Due to symmetry in operation in the identical core structures, the PQVCO is likely to exhibit bi-modal oscillation [38]. But in presence of mismatch in the tank circuits or coupling devices one of the modes prevails and leads to oscillation with either 90° or $-90°$ phase shifted outputs. The ambiguity in phase is resolved by introducing ring oscillator structures [2,39] at the output nodes. As shown in Figure 2b, two ring oscillators are provided and either of them can be enabled to direct the oscillation in either leading or lagging phases. The resulting waveforms of $V_{A+}, V_{A-}, V_{B+}, V_{B-}$ based on different ring oscillator configurations are shown in Figure 2c.

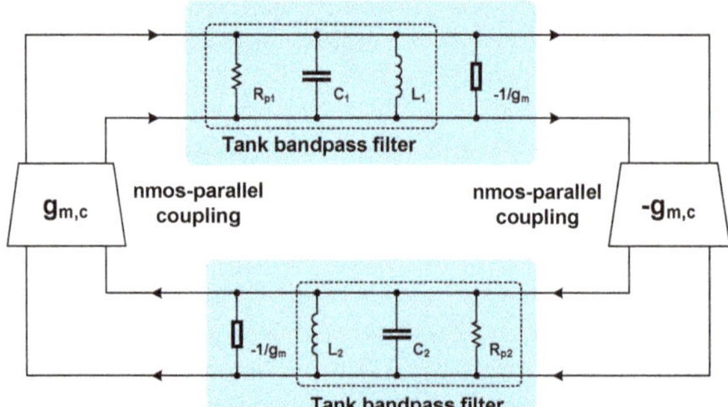

Figure 3. A simplified block diagram of the PQVCO where two oscillator cores are coupled with each other through transconductance ($g_{m,c}$) circuit.

Figure 4 shows a schematic of the SQVCO implemented for the TID experiment based on the architecture proposed in [39]. As elaborated in Figure 4, the two oscillator cores are identical in nature and the common-mode nodes are coupled with each other at 180° out-of-phase through a center-tap inductor. Ideally, in a differential configuration, the odd harmonics of the oscillator circulate through the switching core MOS devices, and higher order even harmonics appear at the common-mode node of the oscillator core. As per the super-harmonic injection locking mechanism [39], the second harmonics of each oscillator are mutually coupled with each other at 180° out-of-phase, and this leads to quadrature phase differences at the fundamental frequency component. Here, in this architecture a center-tap inductor is used in series with the bias circuit and each half is configured to resonate with the parasitic capacitors at twice the fundamental frequency. The second harmonic at the common-mode nodes are coupled magnetically through each half of the inductor and maintain a 180° phase difference between each other. A layout representation of the center-tap inductor working as the coupling transformer is shown in the bottom-left corner of Figure 4. The inductor is implemented in top-metal with dummy metal fillings and guard rings to prevent substrate coupling.

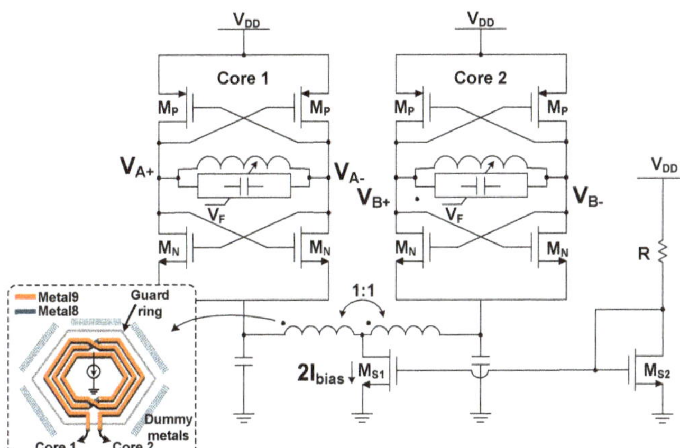

Figure 4. A schematic representation of the super-harmonic coupled quadrature LC-tank VCO (SQVCO).

Unlike the PQVCOs, the absence of coupling through active devices in this architecture breaks the trade-off between quadrature accuracy and phase-noise contribution. Even though the sizes of the core MOS devices are identical in the PQVCO and SQVCO and both are driven by equal bias currents, the phase-noise performance is better in the SQVCO as compared to PQVCO. The coupling inductor acts as a tail-noise filter [40]. It generates a high impedance at twice the fundamental frequency at resonance and therefore reduces flicker noise up-conversion and thermal noise down-conversion around the oscillation frequency.

For better insights in the operation, a simplified block diagram of the SQVCO circuit components [2] are shown in Figure 5. The LC-tank components (R_p, C, L) act as a band-pass filter centered at resonance frequency where the tank losses are compensated by the $-1/g_m$ cell created by the complementary PMOS-NMOS pairs (M_P, M_N). The common-mode node at the oscillator core acts as frequency doubler and injects the signal to the other with an added phase shift of 180°. The injected out-of-phase second harmonic produces a quadrature phase shift while mixing with the fundamental frequency component. Here, in this architecture the quadrature accuracy largely depends on the tail node impedance and resonating signal strength at the tail inductor [39]. Similar to the PQVCO, two ring oscillators are provided at the output nodes as shown in Figure 2b to resolve the quadrature phase ambiguity (lead or lag). In addition, the core PMOS-NMOS switching pairs and the bias circuit of both types of QVCOs: PQVCO and SQVCO, are implemented with p-type and n-type guard-rings to prevent single-event-latchup (SEL) induced failures.

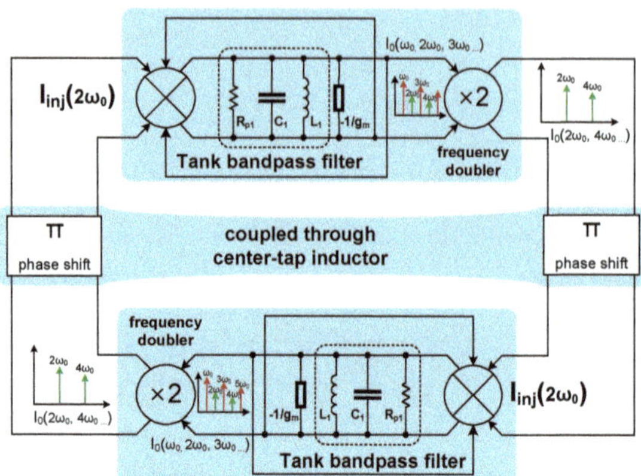

Figure 5. A simplified block diagram of the SQVCO where two oscillator cores coupled with each other through a center-tap inductor.

The off-chip driving strength of the differential output signal pairs from the oscillators are improved using cascaded CMOS digital buffers with increased driving strength. As shown in Figure 6 Two parallel chains of digital buffers are used in differential configuration with cross-coupled latches connected in between the nodes for phase alignment.

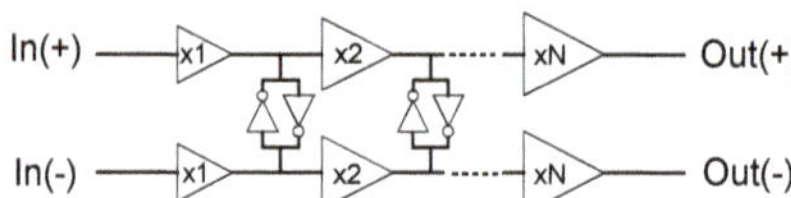

Figure 6. Cascaded CMOS digital buffers used at the oscillator outputs.

3. Experimental Results

The prototypes of the quadrature LC oscillators (PQVCO and SQVCO) are implemented together in a single die using a commercial 9-metal 65 nm CMOS technology. An overview of the test setup used for the radiation assessment of the devices is provided in Figure 7. It consists of 2 printed-circuit-boards (PCBs) and the measurement equipment which are controlled by a raspberry pi 4 based controller board. The device-under-test (DUT) samples with the decoupling capacitors on the core $V_{DD,core}$ and input-output ($V_{DD,IO}$) supplies are wire-bonded to peripheral-component-interconnect-express (PCIe) adapter boards. The micro-photograph of the fabricated die sample is shown in the top-right corner of Figure 7. The buffered quadrature outputs of the oscillators with an approximate frequency range of 2.5 GHz to 2.9 GHz are down-converted to an intermediate frequency (IF) of 500 MHz using an off-chip double balanced active mixer ADL5802. The DUT is connected to the mixer through the PCIe interface, and the mixer is driven by a differential local oscillator (LO) signal (2.4 GHz) generated from low phase noise VCO, ADF4351. The ADL5802 interface PCB with the DUT sample is placed inside the X-ray irradiation chamber. The differential IF quadrature output pairs from the on-board 1:1 balun transformer are brought out of the X-ray chamber to the Keysight DSA91304A oscilloscope using 2 m long sub-miniature version A (SMA) cables for frequency and quadrature phase measurement. The ADL5802 interface PCB and $V_{DD,IO}$ of the DUT are powered from a fixed 5 V and 1.2 V supplies respectively. One of the channels of the Keysight B2902A precision source/measure unit is used to provide 1.2 V to the $V_{DD,core}$ and measure the current consumed. While the other channel is used to tune the varactor input voltage during the measurement. Additionally a PN9000 phase noise analyzer is utilized to measure the output phase noise of the free running quadrature oscillator.

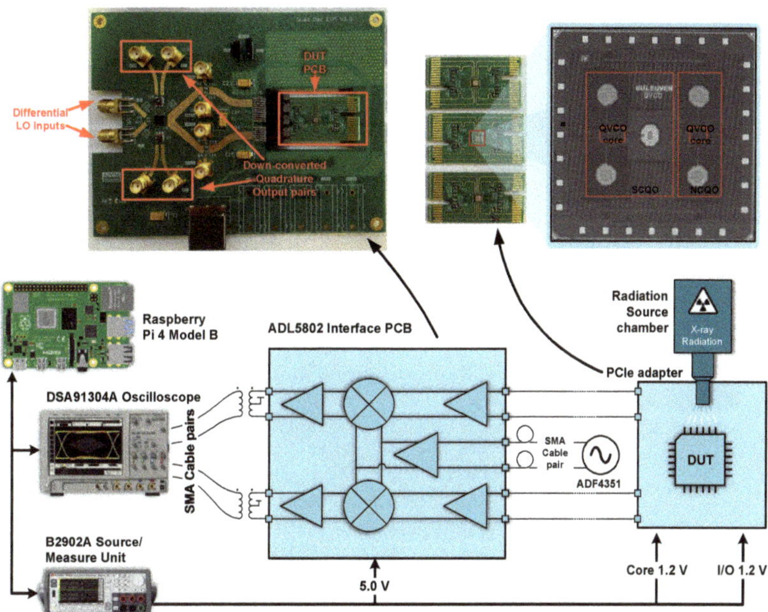

Figure 7. Overview of test setup for radiation assessment of QVCOs and the micrograph of the die (2 mm × 2 mm) in the top right corner.

The radiation assessment of the DUT is conducted under bias condition at room temperature. While testing, the DUT is placed at the center of the incident X-ray beam having a diameter of approximately 3 cm. The X-ray beam is generated from a 40 keV, 40 mA W-tube from Seifert and resulted in a dose rate of 36.78 krad/min. Before irradiation, the dose rate of the X-ray beam is calibrated using a PIN diode-based dose sensor. During irradiation, the various performance metrics (frequency, quadrature phase, core power) of the DUT are measured repeatedly up to a TID level of 100 Mrad (SiO$_2$).

The percentage variations of the measured output frequencies (f_{max}, f_{center} and f_{min}) of PQVCO and SQVCO are shown in Figure 8a and Figure 8b respectively. Before irradiation, the frequency of the PQVCO ranges from 2.524 GHz to 2.786 GHz with a TR of 9.86% and the frequency of the SQVCO ranges from 2.635 GHz to 2.908 GHz with a TR of 9.88%. Similar to the TID experiments [28,29], the oscillation frequency of both the oscillators gradually increases with respect to TID and an increment of approximately 0.5% can be observed at the center frequency. The increase in the frequency can be accounted for primarily due to the decrease in the transconductances (g_{mP}, g_{mN}) of the cross-coupled PMOS-NMOS pair with respect to TID [37]. As elaborated in [41], the frequency of oscillation can be expressed as,

$$f_{OSC}^2 \approx f_0^2\left(1 - \frac{1}{Q_L^2} + \frac{1}{Q_C^2} + \frac{C_{CM}}{C_{tank}}\left(1 - \frac{1}{g_m^2|Z_{s,2}|^2}\right)\right) \tag{1}$$

where f_o is the tank resonance frequency, Q_L is the inductor quality (Q) factor, Q_C is the capacitor Q factor, C_{CM} is the capacitance at the common-mode biasing node, C_{tank} is the tank-capacitance, g_m is effective transconductance of the cross-coupled pair, and $Z_{s,2}$ is the effective impedance at the common-mode biasing mode at the second harmonic of the oscillation frequency.

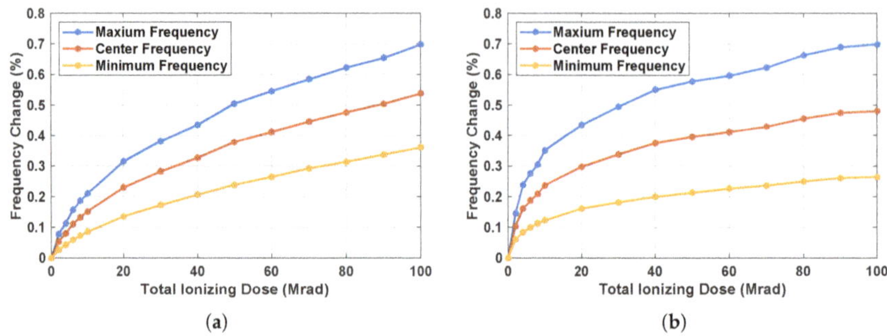

Figure 8. Percentage frequency variations of (**a**) PQVCO and (**b**) SQVCO measured with respect to the TID of the order of 100 Mrad (SiO$_2$).

In the oscillators (PQVCO, SQVCO), the NMOS varactors are ac-coupled and therefore are not affected by the level of output common-mode voltages. As the varactor voltage is tuned from 0 to V_{DD}, the varactors' gate voltage is always greater than the source and drain voltage and so they completely operate in accumulation-mode. The effects of oxide-trap charges and interface trap charges on the varactors with respect to bias voltage are analyzed in detail in [42]. During irradiation, the radiation induced negative charges (oxide-trap and interface trap) result into reduction of the width of the depletion region, which in turn increases the capacitance of the varactors [29,43]. This radiation-induced effect is counteracting previous g_m induced increase in frequency. However, as elaborated in Equation (1) the combined effect of the increase in C_{tank} and the decrease in g_m results in an increase of the frequency of oscillation. In the case of f_{max}, the contribution of C_{tank} is less compared to that of f_{min}. For lower values of C_{tank}, the effect of g_m is much more dominant and, therefore, the relative variation of f_{max} is more than f_{min}. In the case of SQVCO, the effective impedance $Z_{s,2}$ at the common-mode biasing node is larger compared to PQVCO because of the resonance of the coupling inductor. Therefore, the increase in the term $1/(g_m^2 |Z_{s,2}|^2)$ is less in SQVCO than PQVCO, which results in a slightly larger relative variation of f_{min} in PQVCO compared to SQVCO. The variations of the TR of the oscillators, which can be expressed as TR $= 100 \times (f_{max} - f_{min})/f_{center}$, are measured with respect to TID and shown in Figure 9a. After 100 Mrad (SiO$_2$) TID, the relative variation of TR of the PQVCO is 3.4% and that of the SQVCO is 4.4%. In spite of having similar relative variations of f_{max} for both the oscillators, the relative variation of TR of the PQVCO is less compared to SQVCO primarily due to larger variations of f_{min} in the PQVCO compared to the SQVCO.

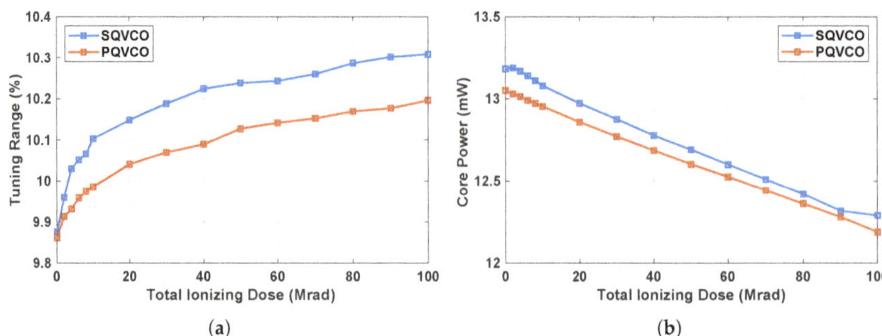

Figure 9. (**a**) Tuning ranges, (**b**) core power consumption of SQVCO and PQVCO measured with respect to the TID of the order of 100 Mrad (SiO$_2$).

The variations of the dissipated core power of the two different oscillators: SQVCO and PQVCO, are shown in Figure 9b. Before irradiation, the core of the PQVCO consumes a power of 13 mW, whereas the core of the SQVCO consumes a power of 13.2 mW. The dissipated core power reduces with respect to TID and after 100 Mrad (SiO$_2$), the core power changes around 7% for both oscillators. NMOS pairs in current-mirror formation with resistive pull-up are used to provide the bias currents to the core of the oscillators. While exposed to radiation, the threshold voltage of the NMOS devices gradually decreases [37], which in turn results in an increased overdrive voltage and a decrease in the current through the resistive pull-up path. Due to mirroring action, the core bias currents also undergo similar reductions. The reduction in bias current also corroborates the fact of increment in oscillation frequency with respect to radiation. As elaborated in [41], the relationship between the oscillation frequency and the bias current is such that when the oscillator is biased at the edge of current-limited region using large bias current, a reduction in bias current shows a slight increase in frequency.

The variations of the absolute Q-phase error of the two oscillators are measured with respect to radiation up to a level of 100 Mrad (SiO$_2$). Before irradiation, the Q-phase error is larger (10°) in the SQVCO compared to the Q-phase error (5.9°) of the PQVCO. The Q-phase error largely depends on the mismatch between the tank resonance of the mutually coupled oscillator cores and the coupling strength of the I/Q phases. The variation of absolute Q-phase error of the two oscillators with respect to the mismatch of C_{tank} based on post-layout simulation results are shown in Figure 10b. With increasing mismatch between C_{tank} of the oscillator cores, the Q-phase error decreases. In the case of the PQVCO, the change in phase error is much more rapid with respect to the mismatch for lower values of the coupling factor (K). During the radiation experiment, the capacitance of the NMOS varactor with respect to the tuning voltage increases with respect to the radiation dose [29,43]. The increasing capacitances eventually reduce the relative mismatch between the C_{tank}. The effect can be seen with respect to radiation as the Q-phase error tends to reduce for both the oscillators as depicted in Figure 10b. As found from the simulation (Figure 10b), increasing the coupling strength (K) diminishes the effects of mismatches.

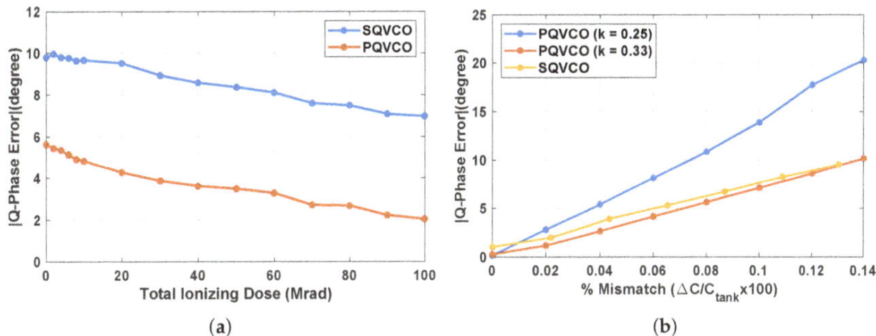

Figure 10. (**a**) Variations of Q-phase error measured with respect to TID of 100 Mrad (SiO$_2$), (**b**) simulated Q-phase error for the SQVCO and PQVCO with respect to different relative mismatch in C_{tank}.

The phase noise characteristics of the PQVCO and the SQVCO measured at frequency f_{center} are shown in Figure 11a and Figure 11b respectively. The blue lines show the phase noise contributions measured prior to radiation exposure and the red ones show the phase noise observed after 100 Mrad (SiO$_2$). As can be observed in Figure 11a,b the phase noise characteristics in the $1/f^2$ region (>300 kHz offset) do not undergo much variation. This is counter intuitive with 7% reduction in the bias current. Here, the oscillators are biased at the edge of the current-limited region to produce full swing outputs and achieve optimum phase noise performance. In this region of operation, the changing bias current has minimal effect on the phase noise characteristics [44] of the oscillators. Although, the thermal noise contribution of the cross-coupled core devices are supposed to increase with respect to

TID [45], the overall phase noise characteristics in the $1/f^2$ region does not change much. This is primarily due to the wider device sizes as used in the cross-coupled pairs which reduce the thermal noise contributions in the phase noise characteristics. However, a close observation can reveal the close-in phase noise in the $1/f^3$ region (<300 kHz offset) increases slightly in the case of the PQVCO compared to the SQVCO. This is due to the extra pair of NMOS devices used for coupling the I and Q phases. The sizes of the devices are less compared to the core devices (coupling factor K = 0.25), and therefore contribute more 1/f noise. Overall, the phase noise characteristic of the SQVCO is slightly better than for the PQVCO due to the tail-noise filter [40] in the form of the coupling inductor present in the SQVCO.

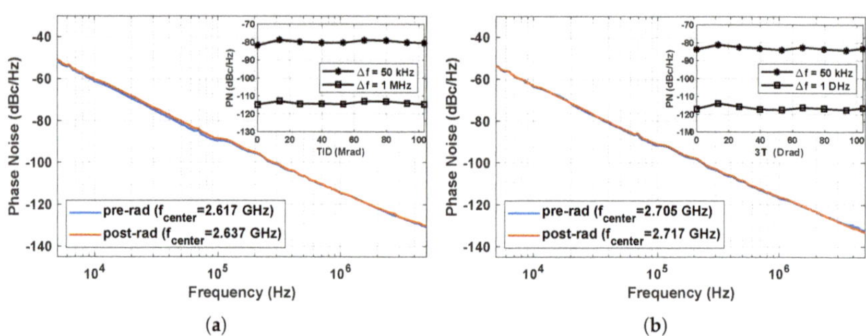

Figure 11. Phase noise measured at f_{center} before (pre-rad) and after (post-rad) radiation exposure of the order of 100 Mrad (SiO_2) from the outputs of the (**a**) PQVCO and (**b**) SQVCO.

Table 1 presents a performance comparison of the implemented QVCOs with respect to previously reported VCOs [28,29,46,47] and QVCOs [33] which are studied under radiation (TID) exposure. Similar to [28], the implemented QVCOs in this work are targeted to operate in the s-band. The implemented QVCOs consume double power in comparison with [28] as two oscillator cores are coupled together to generate quadrature phases. Based on open loop phase noise characteristics, the implemented QVCOs achieve more than 3 dB better phase noise at 1 MHz offset. However, [28] has a similar rate of variations of the frequency with respect to TID level when compared to the implemented QVCOs. The oscillator in [47] has combined small varactors with large digitally switchable capacitor-banks (64 units) and therefore could achieve better phase noise performance compared to the other reported designs. Prior to this work as published in the literature to date, ref. [33] has been the sole instance of TID study performed on QVCOs. Therefore, the results are included in the comparison, even though it is implemented using a 32 nm CMOS SOI technology which is different from the one used in this work. Although TID induced variations tend to reduce with smaller feature sizes, the QVCOs reported in [33] show more frequency variation and less TID tolerance compared to the QVCOs implemented in this work.

Table 1. Performance comparison with respect to previously published VCOs and QVCOs which are tested under radiation (TID).

Reference	TNS'17 [28]	TNS'18 [36]	TNS'18 [46]	TCASI'19 [47]	TNS'21 [29]	TNS'17 [33]	This Work	
Technology (nm)	65	65	65	65	65	32	65	
Type	VCO	VCO	VCO	VCO	VCO	QVCO	QVCO [†]	QVCO [‡]
Oscillator Area (mm^2)	-	-	0.124	-	0.061	0.0484	0.458	0.367
Frequency (GHz)	2.2–3.2	2.5–2.65	4.8–6.0	4.9–5.2	5.4–6.8	20.1–20.7	2.6–2.9	2.5–2.8
Tuning Range (%)	30 **	5.8	4	5.9 **	23	3	9.9	
Phase Noise [§§] @1 MHz (dBc/Hz)	−110	−118	-	−122	−100	−99	−119	−115
VCO Gain (MHz/V)	240 **	-	1850	100 **	225	610	273	262
Power (mW)	6	1.8	18	34	2.85	12.8	13.2	13
FoM [§] (dBc/Hz)	−171	−188.7	-	−180	−171.4	−176	−176.4	−172.2
Frequency Change (%)	3.5	-	-	3.2	2.54	1.4	0.7	
TID Tolerance	600	-	250	350	1000	0.5	100	

[§] FoM = Phase Noise@$\Delta f - 20\log(\frac{f_o}{\Delta f}) + 10\log(\frac{Power}{1\,\mathrm{mW}})$. [†] SQVCO, [‡] PQVCO. [§§] Open-loop phase noise, ** Varactors and digitally-controlled capacitor-banks used.

4. Conclusions

The performance study and evaluation of radiation induced effects on the prototypes of the QVCOs: PQVCO and SQVCO are presented in this article. It shall help to understand the vulnerabilities of the implementation and thereby improve the design for future implementations. The samples of the prototypes are tested under X-ray radiation up to a level of 100 Mrad (SiO$_2$). Prior to radiation exposure, the measured frequency of the PQVCO ranges from 2.524 GHz to 2.786 GHz with a TR of 9.86% and the measured frequency of the SQVCO ranges from 2.635 GHz to 2.908 GHz with a TR of 9.88%. The overall variations of the frequencies of the oscillators are less than 1% and change in TR is less than 5% after the radiation exposure, which makes them suitable to be integrated inside phase-locked-loops (PLLs) and half-rate CDRs. Although the bias current reduces by 7% after 100 Mrad (SiO$_2$) irradiation, it has little contribution on the phase noise performance of the QVCOs. The most vulnerable performance metric is the Q-phase error of the oscillators. Here, in this experiment the decreasing relative mismatch between the C_{tank} improves the Q-phase error. In the case of the oscillator SQVCO the Q-phase error improves from 10° to 7° and in the case of the oscillators PQVCO, it improves from 5.9° to 2°. However, for applications involving much higher frequencies, the value of C_{tank} used would be less, which may worsen the radiation induced variations in Q-phase. The solution would be to increase the coupling strength in the PQVCO to minimize the mismatch induced variations at the expense of phase noise performance. In the case of SQVCO, extra effort is needed to produce matching layouts of the oscillator cores for reducing the mismatch.

Author Contributions: Conceptualization, A.K., V.D.S. and P.L.; Data curation, A.K.; Investigation, A.K.; Methodology, A.K., V.D.S. and P.L.; Supervision, V.D.S. and P.L.; Writing—original draft, A.K.; Writing—review and editing, A.K., V.D.S. and P.L. All authors have read and agreed to the published version of the manuscript.

Funding: This work is funded by the European's Union Horizon 2020 Research and Innovation Programme under grant agreement number 721624 'RADSAGA'.

Institutional Review Board Statement: Not applicable.

Informed Consent Statement: Not applicable.

Data Availability Statement: Not applicable.

Conflicts of Interest: The authors declare no conflict of interest. The funders had no role in the design of the study; in the collection, analyses, or interpretation of data; in the writing of the manuscript, or in the decision to publish the results.

References

1. Musa, A.; Murakami, R.; Sato, T.; Chaivipas, W.; Okada, K.; Matsuzawa, A. A Low Phase Noise Quadrature Injection Locked Frequency Synthesizer for MM-Wave Applications. *IEEE J. Solid-State Circuits* **2011**, *46*, 2635–2649. [CrossRef]
2. Chamas, I.R.; Raman, S. Analysis and Design of a CMOS Phase-Tunable Injection-Coupled LC Quadrature VCO (PTIC-QVCO). *IEEE J. Solid-State Circuits* **2009**, *44*, 784–796. [CrossRef]
3. Eroglu, A. Non-Invasive Quadrature Modulator Balancing Method to Optimize Image Band Rejection. *IEEE Trans. Circuits Syst. I Regul. Pap.* **2014**, *61*, 600–612. [CrossRef]
4. Nikoofard, A.; Kananian, S.; Fotowat-Ahmady, A. A fully analog calibration technique for phase and gain mismatches in image-reject receivers. *AEU-Int. J. Electron. Commun.* **2015**, *69*, 823–835. [CrossRef]
5. Lee, J.; Razavi, B. A 40-Gb/s clock and data recovery circuit in 0.18-μm CMOS technology. *IEEE J. Solid-State Circuits* **2003**, *38*, 2181–2190. [CrossRef]
6. Mazzanti, A.; Svelto, F.; Andreani, P. On the amplitude and phase errors of quadrature LC-tank CMOS oscillators. *IEEE J. Solid-State Circuits* **2006**, *41*, 1305–1313. [CrossRef]
7. Natarajan, A.; Komijani, A.; Guan, X.; Babakhani, A.; Hajimiri, A. A 77-GHz Phased-Array Transceiver with On-Chip Antennas in Silicon: Transmitter and Local LO-Path Phase Shifting. *IEEE J. Solid-State Circuits* **2006**, *41*, 2807–2819. [CrossRef]
8. Scheir, K.; Bronckers, S.; Borremans, J.; Wambacq, P.; Rolain, Y. A 52 GHz Phased-Array Receiver Front-End in 90 nm Digital CMOS. *IEEE J. Solid-State Circuits* **2008**, *43*, 2651–2659. [CrossRef]
9. Nikoofard, A.; Kananian, S.; Fotowat-Ahmady, A. Off-Resonance Oscillation, Phase Retention, and Orthogonality Modeling in Quadrature Oscillators. *IEEE Trans. Circuits Syst. I Regul. Pap.* **2016**, *63*, 883–894. [CrossRef]
10. Zhang, L.; Kuo, N.C.; Niknejad, A.M. A 37.5–45 GHz Superharmonic-Coupled QVCO with Tunable Phase Accuracy in 28 nm CMOS. *IEEE J. Solid-State Circuits* **2019**, *54*, 2754–2764. [CrossRef]
11. Oliveira, L.B.; Fernandes, J.R.; Filanovsky, I.M.; Verhoeven, C.J.; Silva, M.M. *Analysis and Design of Quadrature Oscillators*; Springer Science & Business Media: Berlin/Heidelberg, Germany, 2008.
12. Koo, J.; Kim, B.; Park, H.J.; Sim, J.Y. A Quadrature RC Oscillator with Noise Reduction by Voltage Swing Control. *IEEE Trans. Circuits Syst. I Regul. Pap.* **2019**, *66*, 3077–3088. [CrossRef]
13. Dai, L.; Harjani, R. A low-phase-noise CMOS ring oscillator with differential control and quadrature outputs. In Proceedings of the 14th Annual IEEE International ASIC/SOC Conference (IEEE Cat. No.01TH8558), Arlington, VA, USA, 12–15 September 2001; pp. 134–138. [CrossRef]
14. Pankratz, E.J.; Sanchez-Sinencio, E. Multiloop High-Power-Supply-Rejection Quadrature Ring Oscillator. *IEEE J. Solid-State Circuits* **2012**, *47*, 2033–2048. [CrossRef]
15. Behbahani, F.; Kishigami, Y.; Leete, J.; Abidi, A. CMOS mixers and polyphase filters for large image rejection. *IEEE J. Solid-State Circuits* **2001**, *36*, 873–887. [CrossRef]
16. Sah, S.P.; Yu, X.; Heo, D. Design and Analysis of a Wideband 15–35-GHz Quadrature Phase Shifter with Inductive Loading. *IEEE Trans. Microw. Theory Tech.* **2013**, *61*, 3024–3033. [CrossRef]
17. Lee, T.H. *The Design of CMOS Radio-Frequency Integrated Circuits*; Cambridge University Press: Cambridge, UK, 2003.
18. Andreani, P.; Bonfanti, A.; Romano, L.; Samori, C. Analysis and design of a 1.8-GHz CMOS LC quadrature VCO. *IEEE J. Solid-State Circuits* **2002**, *37*, 1737–1747. [CrossRef]
19. Chang, H.Y.; Chiu, Y.T. *K*-Band CMOS Differential and Quadrature Voltage-Controlled Oscillators for Low Phase-Noise and Low-Power Applications. *IEEE Trans. Microw. Theory Tech.* **2012**, *60*, 46–59. [CrossRef]
20. Chang, H.Y.; Lin, C.H.; Liu, Y.C.; Yeh, Y.L.; Chen, K.; Wu, S.H. 65-nm CMOS Dual-Gate Device for Ka-Band Broadband Low-Noise Amplifier and High-Accuracy Quadrature Voltage-Controlled Oscillator. *IEEE Trans. Microw. Theory Tech.* **2013**, *61*, 2402–2413. [CrossRef]
21. Decanis, U.; Ghilioni, A.; Monaco, E.; Mazzanti, A.; Svelto, F. A Low-Noise Quadrature VCO Based on Magnetically Coupled Resonators and a Wideband Frequency Divider at Millimeter Waves. *IEEE J. Solid-State Circuits* **2011**, *46*, 2943–2955. [CrossRef]
22. Kuo, N.C.; Chien, J.C.; Niknejad, A.M. Design and Analysis on Bidirectionally and Passively Coupled QVCO With Nonlinear Coupler. *IEEE Trans. Microw. Theory Tech.* **2015**, *63*, 1130–1141. [CrossRef]
23. Karmakar, A.; Wang, J.; Prinzie, J.; De Smedt, V.; Leroux, P. A Review of Semiconductor Based Ionising Radiation Sensors Used in Harsh Radiation Environments and Their Applications. *Radiation* **2021**, *1*, 194–217. [CrossRef]
24. Faccio, F.; Michelis, S.; Cornale, D.; Paccagnella, A.; Gerardin, S. Radiation-Induced Short Channel (RISCE) and Narrow Channel (RINCE) Effects in 65 and 130 nm MOSFETs. *IEEE Trans. Nucl. Sci.* **2015**, *62*, 2933–2940. [CrossRef]
25. Borghello, G. Ionizing Radiation Effects in Nanoscale CMOS Technologies Exposed to Ultra-High Doses. Ph.D. Thesis, University of Udine, Udine, Italy, 2019.

26. Borghello, G.; Lerario, E.; Faccio, F.; Koch, H.; Termo, G.; Michelis, S.; Marquez, F.; Palomo, F.; Muñoz, F. Ionizing radiation damage in 65 nm CMOS technology: Influence of geometry, bias and temperature at ultra-high doses. *Microelectron. Reliab.* **2021**, *116*, 114016. [CrossRef]
27. Wang, T.; Wang, K.; Chen, L.; Dinh, A.; Bhuva, B.; Shuler, R. A RHBD LC-Tank Oscillator Design Tolerant to Single-Event Transients. *IEEE Trans. Nucl. Sci.* **2010**, *57*, 3620–3625. [CrossRef]
28. Prinzie, J.; Christiansen, J.; Moreira, P.; Steyaert, M.; Leroux, P. Comparison of a 65 nm CMOS Ring- and LC-Oscillator Based PLL in Terms of TID and SEU Sensitivity. *IEEE Trans. Nucl. Sci.* **2017**, *64*, 245–252. [CrossRef]
29. Monda, D.; Ciarpi, G.; Saponara, S. Design and Verification of a 6.25 GHz LC-Tank VCO Integrated in 65 nm CMOS Technology Operating up to 1 Grad TID. *IEEE Trans. Nucl. Sci.* **2021**, *68*, 2524–2532. [CrossRef]
30. Jagtap, S.; Sharma, D.; Gupta, S. Design of SET tolerant LC oscillators using distributed bias circuitry. *Microelectron. Reliab.* **2015**, *55*, 1537–1541. [CrossRef]
31. Karthigeyan, A.; Radha, S.; Manikandan, E. Single event transient mitigation techniques for a cross-coupled LC oscillator, including a single-event transient hardened CMOS LC-VCO circuit. *IET Circuits Devices Syst.* **2022**, *16*, 178–188. [CrossRef]
32. Jagtap, S.; Anmadwar, S.; Rudrapati, S.; Gupta, S. A Single-Event Transient-Tolerant High-Frequency CMOS Quadrature Phase Oscillator. *IEEE Trans. Nucl. Sci.* **2019**, *66*, 2072–2079. [CrossRef]
33. Loveless, T.D.; Jagannathan, S.; Zhang, E.X.; Fleetwood, D.M.; Kauppila, J.S.; Haeffner, T.D.; Massengill, L.W. Combined Effects of Total Ionizing Dose and Temperature on a K-Band Quadrature LC-Tank VCO in a 32 nm CMOS SOI Technology. *IEEE Trans. Nucl. Sci.* **2017**, *64*, 204–211. [CrossRef]
34. Chamas, I.R.; Raman, S. A Comprehensive Analysis of Quadrature Signal Synthesis in Cross-Coupled RF VCOs. *IEEE Trans. Circuits Syst. I Regul. Pap.* **2007**, *54*, 689–704. [CrossRef]
35. Mirzaei, A.; Heidari, M.E.; Bagheri, R.; Chehrazi, S.; Abidi, A.A. The Quadrature LC Oscillator: A Complete Portrait Based on Injection Locking. *IEEE J. Solid-State Circuits* **2007**, *42*, 1916–1932. [CrossRef]
36. Prinzie, J.; Christiansen, J.; Moreira, P.; Steyaert, M.; Leroux, P. A 2.56-GHz SEU Radiation Hard *LC*-Tank VCO for High-Speed Communication Links in 65-nm CMOS Technology. *IEEE Trans. Nucl. Sci.* **2018**, *65*, 407–412. [CrossRef]
37. Menouni, M.; Barbero, M.; Bompard, F.; Bonacini, S.; Fougeron, D.; Gaglione, R.; Rozanov, A.; Valerio, P.; Wang, A. 1-Grad total dose evaluation of 65 nm CMOS technology for the HL-LHC upgrades. *J. Instrum.* **2015**, *10*, C05009. [CrossRef]
38. Tong, H.; Cheng, S.; Lo, Y.C.; Karsilayan, A.I.; Silva-Martinez, J. An LC Quadrature VCO Using Capacitive Source Degeneration Coupling to Eliminate Bi-Modal Oscillation. *IEEE Trans. Circuits Syst. I Regul. Pap.* **2012**, *59*, 1871–1879. [CrossRef]
39. Gierkink, S.; Levantino, S.; Frye, R.; Samori, C.; Boccuzzi, V. A low-phase-noise 5-GHz CMOS quadrature VCO using superharmonic coupling. *IEEE J. Solid-State Circuits* **2003**, *38*, 1148–1154. [CrossRef]
40. Hegazi, E.; Sjoland, H.; Abidi, A. A filtering technique to lower LC oscillator phase noise. *IEEE J. Solid-State Circuits* **2001**, *36*, 1921–1930. [CrossRef]
41. Wu, T.; Moon, U.K.; Mayaram, K. Dependence of LC VCO oscillation frequency on bias current. In Proceedings of the 2006 IEEE International Symposium on Circuits and Systems (ISCAS), Island of Kos, Greece, 21–24 May 2006; p. 4. [CrossRef]
42. Heiman, F.; Warfield, G. The effects of oxide traps on the MOS capacitance. *IEEE Trans. Electron Devices* **1965**, *12*, 167–178. [CrossRef]
43. Fernández-Martínez, P.; Cortés, I.; Hidalgo, S.; Flores, D.; Palomo, F.R. Simulation of Total Ionising Dose in MOS capacitors. In Proceedings of the 8th Spanish Conference on Electron Devices, CDE'2011, Palma de Mallorca, Spain, 8–11 February 2011; pp. 1–4. [CrossRef]
44. Rael, J.; Abidi, A.A. Physical processes of phase noise in differential LC oscillators. In Proceedings of the IEEE 2000 Custom Integrated Circuits Conference (Cat. No. 00CH37044), Orlando, FL, USA, 24 May 2000; pp. 569–572.
45. Re, V.; Gaioni, L.; Manghisoni, M.; Ratti, L.; Traversi, G. Comprehensive Study of Total Ionizing Dose Damage Mechanisms and Their Effects on Noise Sources in a 90 nm CMOS Technology. *IEEE Trans. Nucl. Sci.* **2008**, *55*, 3272–3279. [CrossRef]
46. Mazza, G.; Panati, S. A Compact, Low Jitter, CMOS 65 nm 4.8–6 GHz Phase-Locked Loop for Applications in HEP Experiments Front-End Electronics. *IEEE Trans. Nucl. Sci.* **2018**, *65*, 1212–1217. [CrossRef]
47. Biereigel, S.; Kulis, S.; Leitao, P.; Francisco, R.; Moreira, P.; Leroux, P.; Prinzie, J. A low noise fault tolerant radiation hardened 2.56 Gbps clock-data recovery circuit with high speed feed forward correction in 65 nm CMOS. *IEEE Trans. Circuits Syst. I Regul. Pap.* **2019**, *67*, 1438–1446. [CrossRef]

 *electronics*

MDPI

Article

Radiation-Tolerant All-Digital PLL/CDR with Varactorless LC DCO in 65 nm CMOS

Stefan Biereigel [1,2,3,*] , Szymon Kulis [1] , Paulo Moreira [1] , Alexander Kölpin [4] , Paul Leroux [2] and Jeffrey Prinzie [2]

1   CERN, The European Center for Nuclear Research, 1211 Meyrin, Switzerland; Szymon.kulis@cern.ch (S.K.); paulo.moreira@cern.ch (P.M.)
2   ESAT-ADVISE Research Lab, KU Leuven University, 3000 Leuven, Belgium; paul.leroux@kuleuven.be (P.L.); jeffrey.prinzie@kuleuven.be (J.P.)
3   Chair of Electronics and Sensor Systems, Brandenburg University of Technology, 03046 Cottbus, Germany
4   Institute for High Frequency Technology, Hamburg University of Technology, 21073 Hamburg, Germany; alexander.koelpin@tuhh.de
*   Correspondence: Stefan.Biereigel@cern.ch

Abstract: This paper presents the first fully integrated radiation-tolerant All-Digital Phase-Locked Loop (PLL) and Clock and Data Recovery (CDR) circuit for wireline communication applications. Several radiation hardening techniques are proposed to achieve state-of-the-art immunity to Single-Event Effects (SEEs) up to 62.5 MeV cm^2 mg^{-1} as well as tolerance to the Total Ionizing Dose (TID) exceeding 1.5 Grad. The LC Digitally Controlled Oscillator (DCO) is implemented without MOS varactors, avoiding the use of a highly SEE sensitive circuit element. The circuit is designed to operate at reference clock frequencies from 40 MHz to 320 MHz or at data rates from 40 Mbps to 320 Mbps and displays a jitter performance of 520 fs with a power dissipation of only 11 mW and an FOM of −235 dB.

Keywords: All-Digital; PLL; CDR; Single-Event Effects; radiation hardening

Citation: Biereigel, S.; Kulis, S.; Moreira, P.; Kölpin, A.; Leroux, P.; Prinzie, J. Radiation-Tolerant All-Digital PLL/CDR with Varactorless LC DCO in 65 nm CMOS. *Electronics* **2021**, *10*, 2741. https://doi.org/10.3390/electronics10222741

Academic Editor: Gianluca Traversi

Received: 11 October 2021
Accepted: 5 November 2021
Published: 10 November 2021

Publisher's Note: MDPI stays neutral with regard to jurisdictional claims in published maps and institutional affiliations.

1. Introduction

Phase-locked loops, including Clock/Data Recovery circuits, play an important role in the reliability of radiation tolerant systems, being typically responsible for providing and conditioning clocks of high spectral purity and stability. With the emergence of ever more stringent timing precision and stability requirements in both High Energy Physics applications [1] as well as communications circuits for space applications [2], PLLs operating in radiation environments need to provide radiation tolerance without sacrificing performance.

Without extensive circuit hardening efforts, conventional PLL architectures have been found to be sensitive to ionizing radiation, both with regards to SEE as well as TID effects. While their SEE sensitivity is the origin of transient but recoverable fault conditions, the sensitivity to TID ultimately limits the application of these circuits in harsh radiation environments, where reliable operation must be sustained up to radiation doses exceeding 1 Grad [3].

All-Digital Phase-Locked Loop (ADPLL) architectures offer a unique opportunity to eliminate many of these sensitivities and therefore improve radiation tolerant PLL circuits in multiple regards: Firstly, most sensitive analog circuits can be replaced with their digital counterparts. While the design and hardening of analog circuit components in advanced technology nodes become more difficult to accomplish, digital circuits can be hardened against SEE systematically using techniques such as Triple Modular Redundancy (TMR) and temporal redundancy [4], while continuing to exploit the performance gain these nodes provide. At the same time, all-digital architectures allow the tight integration of Digital Signal Processing (DSP) algorithms, which may be leveraged to achieve improved locking times or robust circuit calibration capabilities [5]. Being implemented with digital circuits, such functionalities can become intrinsically immune to TID degradation.

In this article, we propose an All-Digital PLL circuit tolerant to SEE and high TID, which is based on a varactorless LC DCO implemented in a 65 nm CMOS technology. We augment the classical ADPLL architecture with the inclusion of SEE protection for all digital circuit components. To accomplish this goal, we take advantage of the availability of systematic SEE hardening techniques for digital circuits. The LC DCO, which constitutes the sole remaining mixed signal circuit component, is implemented without MOS varactors, which we show to be a beneficial implementation in radiation environments. The chosen architecture eliminates conceptual SEE sensitivities found in conventional PLLs, while, at the same time, significantly reducing the impact of TID, enabling the use of PLL circuits in challenging radiation environments. As a case study, the proposed circuit targets common clocking requirements of the High Luminosity Large Hadron Collider (HL-LHC) experiments. These systems typically operate on a reference clock or data stream derived from a 40 MHz bunch crossing clock and often simultaneously require multiple low-jitter clocks up to 1.28 GHz.

The remainder of this paper is organized as follows: Section 2 gives an overview of radiation effects in conventional PLL architectures. Section 3 motivates the application of an All-Digital PLL architecture to systematically mitigate radiation effects, including the proposal of suitable hardening techniques. Section 4 describes the circuit implementation, and Section 5 presents the measurements of the circuit performance and radiation tolerance.

2. Radiation Effects in Conventional PLL Designs

Radiation effects in conventional PLL designs have been the focus of numerous studies in the past. We limit the discussion of radiation effects to circuits based on LC tank Voltage-Controlled Oscillators (VCOs) as this topology is most often chosen for low power and low phase noise circuit implementations. Conventional PLL circuits are comprised of a linear Phase-Frequency Detector (PFD), Charge Pump (CP), Loop Filter (LF), LC VCO and a feedback divider, which closes the feedback loop. Following the general classification of radiation effects in integrated circuits, we can separate their sensitivity to radiation into transient (SEE) and cumulative (TID) effects [6]. SEE in particular can be divided into two relevant classes: Single-Event Upsets (SEUs) refer to events altering the state of memory elements, which remain persistent until overwritten or corrected. On the other hand, Single-Event Transients (SETs) specifically refer to temporary changes in voltages or logic levels that recover without further intervention. Both effects, when not mitigated, can have catastrophic impact on the operation of digital circuits.

First, considering only Single-Event Effects, a number of shortcomings of the conventional PLL architecture have been reported in the past. The PFD containing digital storage elements is generally sensitive to SEUs. Even though an upset in the PFD persists for, at most, one reference clock cycle, it may cause the charge pumps to remain enabled for a significant amount of time, which are otherwise (in steady state operation) only active for a very short period each reference clock cycle. Bit flips can also corrupt the state of the PFD, which results in cycle slips [7,8]. This sensitivity can be largely mitigated by the protection of the PFD using TMR [7]. A similar reasoning can be applied to the feedback divider: Corruption of any of the memory elements in the divider results in a phase jump of the feedback clock. This phase error subsequently needs to be corrected by the feedback loop. Again, a mitigation can be found in TMR implementations of the divider [7,9] or by applying Radiation-Hardened-by-Design (RHBD) techniques, such as Double Interlocked Cell (DICE) implementations for memory elements [10].

Another contributor to SEE sensitivity may be found in the charge pump driving the loop filter [8,11]. Current-based charge pumps specifically suffer from long recovery times after SETs due to their limited rate of charge evacuation, which again results in large accumulated phase and frequency errors. This sensitivity has been mitigated, for example, by the use of a tri-state voltage charge pump architecture, which reduces the number of sensitive nodes and increases the rate of charge sourcing/sinking [12,13] at the cost of poor reference-spurs and process, voltage and temperature (PVT) sensitive static phase errors.

Finally, the LC VCO has been identified as a critical component sensitive to SEEs. Charge collection at the bias current source transistor drain node can be shown to result in a temporary reduction of oscillation amplitude and frequency [14,15]. This sensitivity can be significantly reduced by the addition of capacitance to this node [14]. Another sensitivity is located at the VCO tuning node. Typical MOS varactor tuning architectures have been found to be sensitive to charge collection, as reported in [16]. This sensitivity can be mitigated by adopting a modified tuning topology [9], which, however, has the drawback of introducing an additional pole at the tuning node of the VCO.

With regards to TID, the VCO and CP circuits can be identified as the main contributors to performance degradation. The degradation of the PLL's digital circuits (PFD and feedback divider) has no impact on PLL performance as long as timing requirements (setup/hold) are met. The degradation of the devices forming the VCO amplifier results in a reduction in their transconductance. This reduces the oscillation amplitude, which, through the presence of voltage-dependent capacitances at the oscillator nodes, is converted into a significant change of oscillation frequency [16]. Compensating for this shift in oscillation frequency requires extending the VCO tuning range to support operation at high doses. Furthermore, the reduced transconductance of the VCO amplifier will eventually result in failure to start or sustain stable oscillation. Another concern is the degradation of the active devices in the charge pump. A reduction in their current results in a reduction in loop gain, which reduces the bandwidth and damping of the closed-loop transfer function.

3. Radiation-Tolerant All-Digital PLL and CDR Architecture

The All-Digital PLL architecture offers an opportunity for the systematic elimination of many of the sensitivities that the conventional PLL architecture contains. The general SEE sensitivity of different All-Digital PLL architectures without any applied hardening strategies has been reported in the literature before [17,18]. Two analyses have shown that a bang-bang-ADPLL architecture, which lends itself to implementations of Integer-N PLL and CDR circuits, already exhibits inherent robustness due to its nonlinear phase detector. In such an architecture, the Digital Loop Filter (DLF) presents a critical SEE sensitivity, as SEUs may corrupt the frequency error information stored in the loop filter integrator registers. Prior work addressing SEE sensitivity of ADPLL demonstrated its findings using simulations and fault injection into non-hardened circuit implementations [19], but no radiation-tolerant implementation has been demonstrated experimentally. Possible design trade-offs and implementation aspects of hardening techniques were not addressed, and LC DCOs have been excluded from the discussion as a source of SEE.

As outlined earlier, digital circuits can generally be reliably protected against SEEs using redundancy techniques, such as TMR. Taking this into account, we propose a circuit architecture following Figure 1. Depending on the mode of operation, either a D-Flip-Flop (PLL operation) or an Alexander Phase Detector [20] (CDR operation) is used as a single bit phase detector. We propose to only use a single phase detector without any internal redundancy (TMR). A corruption of the phase detector output by SEE will only persist for a single reference clock cycle. Such an error can be tolerated to propagate through the loop, as it does not result in an appreciable phase error of the loop. Since both types of Bang-Bang Phase Detectors (BBPDs) do not offer frequency detection capability, a Frequency Counter (FCNT) is used during acquisition to bring the DCO frequency close to the reference clock frequency before enabling closed-loop control. An Acquisition Finite State Machine (FSM) is implemented for this purpose.

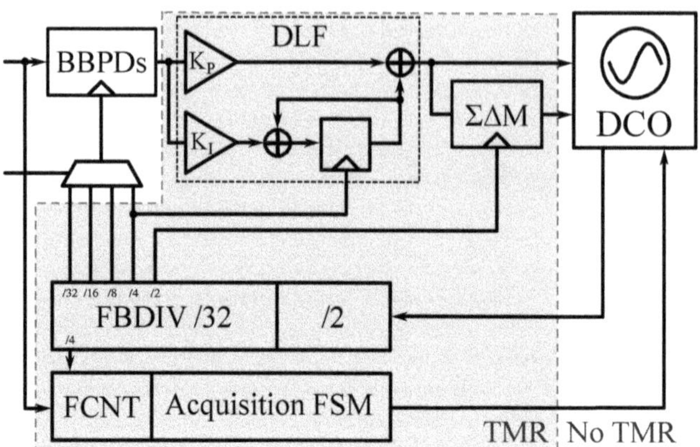

Figure 1. Architecture of the proposed All-Digital PLL/CDR circuit. All synchronous digital loop components are protected using TMR, only the DCO and BBPD remain without redundancy.

The proportional-integral digital loop filter replaces both the charge pump and loop filter circuits present in the conventional PLL architecture. Both the TID-stimulated degradation of the CP currents, as well as the sensitivity of the CP output switches to SEE discussed above, can be systematically removed by replacing the charge pumps with a digital filter. The SEE sensitivity of the DLF identified in [17] is mitigated by applying TMR protection to this circuit. A TMR protection scheme is also used in the $\Sigma\Delta$-modulator, which is employed to improve frequency resolution and shift tuning quantization noise to high frequencies, where it is effectively filtered by the DCO transfer function [21]. A second-order Sigma-Delta Modulator ($\Sigma\Delta M$) is used, providing sufficient self-dithering during closed-loop operation to avoid the generation of spurious tones.

To achieve good phase noise performance, an LC tank oscillator is chosen as the DCO topology. Addressing both the sensitivity to SET and TID of the conventional LC VCO, the LC DCO is implemented without using MOS varactors. Other reported DCO designs utilize small MOS varactors with nonlinear control [5]; however, varactorless implementation has also been reported before [22,23]. By exclusively utilizing Metal-Oxide-Metal (MOM) capacitors fabricated using the highly controlled metallization processes available in CMOS technologies, sufficient tuning range, linearity and frequency resolution can be achieved. Avoiding the use of MOS varactors eliminates their inherent susceptibility to charge collection and therefore a potential SEE sensitivity of the circuit. At the same time, the capacitance of MOM structures offers reduced voltage-dependence, which reduces supply voltage sensitivity and limits TID-stimulated oscillator frequency shift. The DCO can additionally benefit from the omission of linearly controlled varactors as it eliminates a significant contributor to oscillator phase noise through the AM-PM conversion process of the voltage noise present on their tuning node [24], for instance, from the tail current source or supply voltage.

Finally, a synchronous binary counter divider is used in the PLL feedback path. The synchronous implementation lends itself to hardening against SEE using TMR in a similar way as the other digital loop components. To the advantage of our application, it also provides access to intermediate frequencies of the divider, which can be used for clocking the digital loop components (DLF and DCO $\Sigma\Delta$-modulator) and allows flexible selection of the reference clock frequency and data rate.

4. Implementation

The proposed circuit was manufactured in a commercial 65 nm CMOS process with a nominal supply voltage of 1.2 V. The circuit occupies a total area of 0.28 mm², of which only

0.03 mm^2 (11%) is active area occupied by the digital loop components. A photomicrograph of the manufactured circuit is shown in Figure 2.

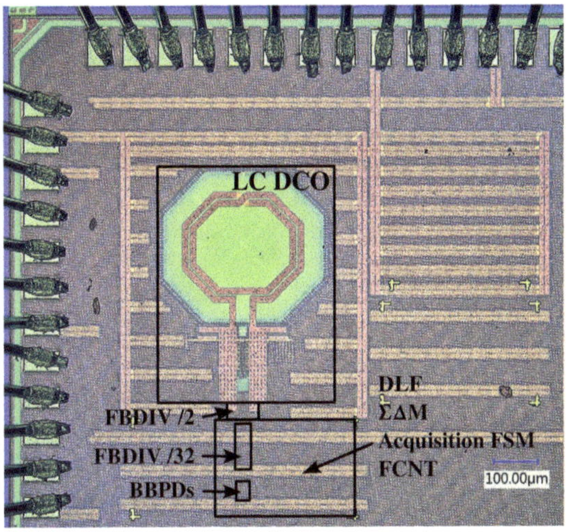

Figure 2. Photomicrograph of the All-Digital PLL/CDR circuit. The LC DCO occupies the majority of the circuit area. Critical digital components are implemented as small cells, while the remaining logic implemented in a sea-of-logic fashion.

4.1. LC DCO

To synthesize output frequencies between 40 MHz and 1280 MHz, an LC DCO operating at 2.56 GHz was designed. When considering the overall area, inductor quality factor and power consumption of the DCO and clock divider, this center frequency provides a suitable trade-off. The 2.56 GHz center frequency is low enough to allow the implementation of all dividers in the feedback path using CMOS circuits. This reduces power consumption and complexity compared to the Current-Mode Logic (CML) implementations often used at higher frequencies, such as, for the 5.12 GHz oscillator presented in [9]. A differential oscillator topology using cross-coupled NMOS and PMOS pairs is used, as shown conceptually in Figure 3. Bias current is provided by an NMOS tail current source. Because of the limited voltage swing this architecture produces, thin-oxide transistors can be used for the cross-coupled pair without reliability concerns and with the added benefit of their improved TID tolerance compared to the thick-oxide devices [25] required for larger oscillation amplitudes. Similar to DCO designs, such as [21], a coarse capacitor bank is used to cover PVT-related differences of the oscillator center frequency. An acquisition bank with improved resolution allows centering the fine tracking bank, which is then used for closed-loop operation. The PVT and acquisition bank settings are established during start-up using the FCNT acquisition FSM to remove large DCO frequency errors.

To cover the expected process and TID-related variations, the DCO is designed for a tuning range of 20%. This range is covered by the coarse PVT bank (C_{PVT}), which is segmented into four binary weighted LSB cells and four additional thermometric MSB cells. Each cell is implemented using MOM capacitors and NMOS switches, with the smallest binary cell implementing a 12.5 fF capacitance difference corresponding to a frequency step size of 6.5 MHz at 2.56 GHz. A smaller acquisition bank (C_{ACQ}) offers additional frequency tuning capabilities. To ensure optimal frequency centering of the small tracking bank (C_{TRK}) used for closed-loop control, the acquisition bank is implemented using 64 thermometric cells. Each unit cell adds 600 aF of capacitance to the tank when enabled, providing a frequency step size of 320 kHz. Both the PVT and the acquisition bank cells are implemented using the bottom-pinning switch architecture presented in [26]. A circuit schematic is shown in Figure 4. Foundry MOM capacitors, C_{MOM}, and NMOS switch devices, M_{SW}, are used.

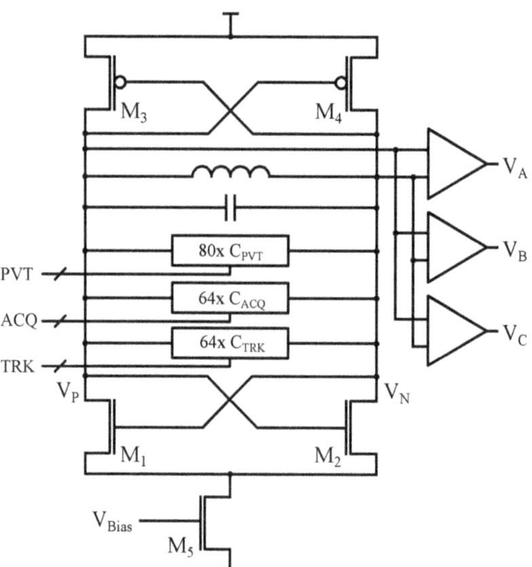

Figure 3. Schematic of the digitally controlled LC oscillator with redundant output buffers. Its frequency is controlled using PVT, acquisition (ACQ) and tracking (TRK) banks. Only the tracking bank is used for closed-loop control.

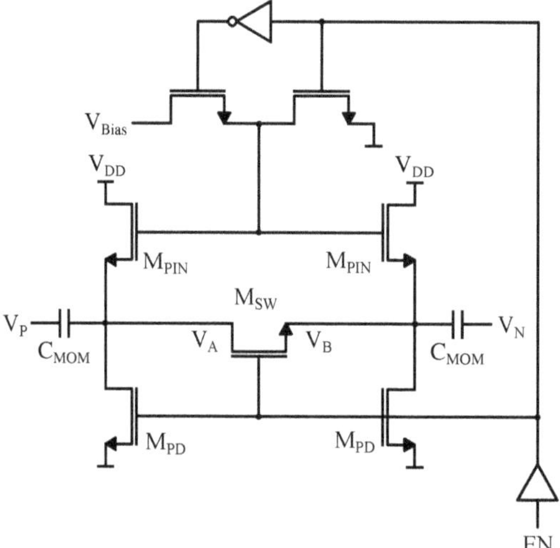

Figure 4. Digitally switched capacitor cells with bottom-pinning biasing [26]. When the main switch transistor M_{SW} is turned off, the NMOS pinning devices M_{PIN} fix the minimum of the oscillation voltage waveform close to the negative supply potential.

To avoid the degradation of the DCO phase noise and to minimize the production of spurious spectral components during closed-loop operation, the tracking bank needs to provide both fine-frequency resolution and high linearity. A custom MOM finger capacitor unit cell is utilized for frequency control. The implementation of this cell is shown in Figure 5. It is composed of four parallel minimum-width metal fingers, stacked across two metallization layers. A grounded poly-silicon shield is implemented below the cell to minimize substrate noise coupling. The two outer metal fingers are connected to the tank oscillation nodes. To digitally modulate the capacitance of this cell, the inner two fingers can be shorted electrically using an NMOS switch, removing a single finger-to-finger capacitance from the tank. Using this arrangement, a capacitance difference of 66 aF can be realized, which provides a 35 kHz frequency resolution of the DCO. This tuning resolution is sufficient considering the phase noise performance of the oscillator and the adopted $\Sigma\Delta$M configuration. The high linearity of the tracking bank is achieved by 64 unit cells placed in a regular and closely matched layout. The capacitor configuration was designed using 2.5D electromagnetic simulations.

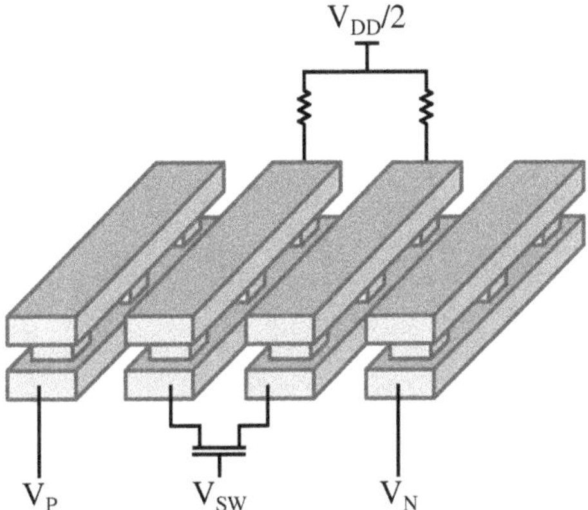

Figure 5. The implementation of the tracking bank custom MOM capacitor cells. Digital control of the cell capacitance is implemented by electrically connecting the innermost two fingers using an NMOS switch.

The DCO is equipped with three identical output buffers, which allows for hardening the remainder of the circuit against SEE using full TMR. These buffers also implement the conversion from differential to single-ended signalling.

4.2. Divide-by-Two Prescaler

Co-integrated with the redundant output buffers and differential to single-ended conversion is a divide-by-two True Single Phase Clock (TSPC) prescaler circuit. To mitigate SEE susceptibility, the prescaler is protected by full TMR, including triplicated voters in the feedback path. To allow sustaining operation at high levels of TID, the prescaler is implemented using a custom high-speed cell library utilizing Enclosed Layout Transistor (ELT) devices [27]. Since the critical path of this prescaler consists only of a single majority voter and the D-Flip-Flop setup time, the prescaler retains significant margins for TID degradation. Three single-ended 1.28 GHz clock signals are provided at the prescaler output.

4.3. Clocking and Clock Distribution

All digital loop components are clocked using frequencies generated by the Feedback Divider (FBDIV). As this divider is protected using TMR, it also provides three replicas of each clock frequency, which allows implementing a full TMR scheme with triplicated clocks. Full triplicating of the clock signals is necessary to mitigate the simultaneous corruption of redundant memory elements by SET affecting their common clock tree, which has catastrophic consequences when affecting the digital loop filter registers, for example.

Clock multiplexing is provided to select the operation frequency of the BBPDs. These can be operated at frequencies of 40, 80, 160 and 320 MHz. The DLF and $\Sigma\Delta$M are clocked at 320 MHz and 640 MHz, respectively. Clock gating functionality allows the operation of these components at lower frequencies depending on the chosen reference frequency. An additional output clock distribution network provides three independent outputs clocks, providing either 40 MHz, 80 MHz, 160 MHz, 320 MHz, 640 MHz or 1280 MHz regardless of the chosen reference clock frequency.

4.4. Digital Loop

All digital loop components (FBDIV, BBPD, DLF and $\Sigma\Delta$M) are implemented using static CMOS logic. With the exception of the divide-by-two prescaler described above, these components are synthesized from RTL code using foundry-provided digital standard cell libraries utilizing ordinary active devices instead of ELTs. This design approach, in addition to improving technology portability, also facilitates the systematic inclusion of SEE hardening: TMR was inserted at the RTL design stage using the TMRG tool [28]. TMRG offers the automatic insertion of TMR protection to Verilog modules based on a small number of user constraints, which increases design automation and eliminates common sources of mistakes during redundancy insertion. Following synthesis, an automatic place and route methodology was applied for the majority of the loop components. Manual placement and routing was limited to the feedback divider and phase detectors, where it is required to preserve signal integrity. To avoid the corruption of memory elements protected by TMR by a multi-bit SEU, a minimum spacing of 15 μm between registers containing replicated information was enforced during placement.

Control over the loop transfer function is given by programmable loop filter coefficients in the proportional and integral paths. The coefficients K_P and K_I can be configured to powers of two between 2^{-10} to 2^5 times the tracking unit cell size. To prevent a corruption of the DLF and $\Sigma\Delta$M integrator registers by the accumulation of SEUs, majority voting is implemented in all circuit feedback paths.

The loop filter output controls the DCO tracking bank using two separate paths: The integer component of the loop filter output word directly drives thermometric unit cells, while its fractional component is generated by three thermometric bits driven by the $\Sigma\Delta$M. To maximize the tolerance of the digital components to TID, conservative setup and hold timings were utilized during implementation. Based on the reported degradation of digital circuits in the chosen technology [29], setup margins of at least 20% of the clock period were used. This allows a sufficient margin for the logic to slow down due to TID damage. Pessimistic hold margins exceeding the foundry recommendations by a factor of at least two have been used to allow for unequal delay degradation of different clocks with inter-clock timing arcs. Since the filter characteristics are defined by digital words rather than analog voltages or currents, the loop dynamics will not change under TID damage until the circuit fails due to violations of setup/hold constraints.

5. Measurements

Three samples were characterized across a supply voltage range of ±10% at room temperature. A Rohde & Schwarz FSWP8 phase noise analyzer was used for phase noise and integrated jitter measurements, while frequency measurements were performed using a Keysight 53220A frequency counter. CDR jitter tolerance performance was evaluated using an Agilent N4903B bit error rate tester.

5.1. DCO Characterization

Tuning range, linearity and step size of the three DCO tuning banks were characterized together with static power consumption and open loop phase noise.

The coarse tuning provided by the PVT bank was confirmed to be well-centered around the nominal oscillation frequency of 2.56 GHz and provide the expected 20% range. A frequency step size of ~6.5 MHz is available around the center frequency. The open loop phase noise of the DCO was measured at the target center frequency, as shown in Figure 6. The phase noise at offset frequencies above 10 MHz is dominated by the clock distribution network feeding the clock test outputs, which was not optimized for low jitter. The oscillator including its divide-by-two prescaler consumes 7.4 mW while providing a phase noise of −123 dBc/Hz (referred to 2.56 GHz) at a carrier offset frequency of 1 MHz. When accounting for the power dissipation of the prescaler (1.8 mW), this results in an oscillator Figure of Merit (FOM) ($\text{FOM}_{\text{DCO}} = -\mathcal{L}(1\,\text{MHz}) + 20\log_{10}\frac{f_{\text{DCO}}}{1\,\text{MHz}} - 10\log_{10}\frac{P_{\text{DCO}}}{1\,\text{mW}}$) of 183.7 dB.

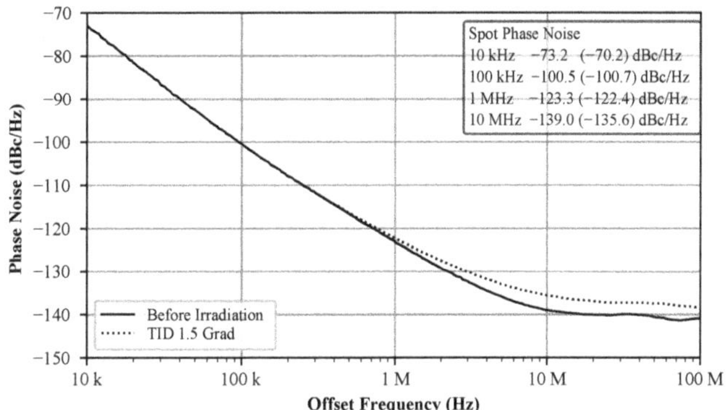

Figure 6. Open loop LC DCO phase noise measurement. Pre- and post-irradiation measurements are shown in solid and dashed lines, respectively. Values in parentheses indicate post-irradiation performance. Noise floor above 10 MHz is limited by clock distribution network.

To validate the custom tracking cell design, the linearity of the tracking bank was further characterized. The very small capacitance of the individual cells demands particular attention during measurements due to the drift of the free-running oscillator. For each tracking bank cell, two measurements were performed, comparing the oscillator frequency at two settings: One of the measurements is performed with all unit cells disabled, while during the second measurement, a fixed number of them are enabled. The obtained frequency difference between these short measurements is stable across long-term drifts of the oscillator, and therefore, multiple measurements can be averaged to reduce the measurement uncertainty. The characterization results can be seen in Figure 7. The thermometric bank offers exceptional differential and integral nonlinearity of better than worst case 0.04 LSB across the full range and a mean frequency step size of 12.7 ppm or 32.5 kHz per cell. This is equivalent to a capacitance difference of 61 aF per cell, which is well in agreement with the electromagnetic simulations.

5.2. PLL and CDR Performance

Closed-loop operation of the ADPLL circuit was tested at all foreseen reference frequencies in PLL and CDR modes of operation across the 1.2 V ± 10% supply voltage range. Typical closed-loop phase noise performance during PLL operation with a 320 MHz reference clock is shown in Figure 8. Integrated random jitter was measured to be 520 fs rms in a 100 Hz to 100 MHz integration band. Jitter tolerance of the CDR circuit operating on 320 Mbit s^{-1} input data is shown in Figure 9. The SONET OC-12 jitter tolerance specification, targeting similar data rates, is included for reference. During operation, the ADPLL circuit dissipates 11 mW of power at the nominal 1.2 V supply voltage. The digital loop components (phase detector, loop filter, $\Sigma\Delta$ modulator and feedback dividers) account for about 50% of the circuit power consumption. This measurement highlights that even with extensive TMR-based hardening against the SEE in place, the DCO still dominates the circuit power consumption in this design. While this property strongly depends on the oscillator design and phase noise requirements of the circuit, it seems fair to assume it will continue to hold in the future for radiation-tolerant designs targeting high-performance applications, especially when the increase of power efficiency for digital circuits in smaller CMOS nodes is taken into account.

A worst case reference spur of −50 dBc was measured at 80 MHz offset from a 640 MHz output clock. This spur was found to originate from power supply coupling between the core logic and the clock distribution network. Switching activity in the digital loop components modulates the clock distribution network supply, which stimulates periodic variations of its propagation delay, creating deterministic jitter. These spurs can be reduced by improving power supply rejection of these buffers or by better supply isolation.

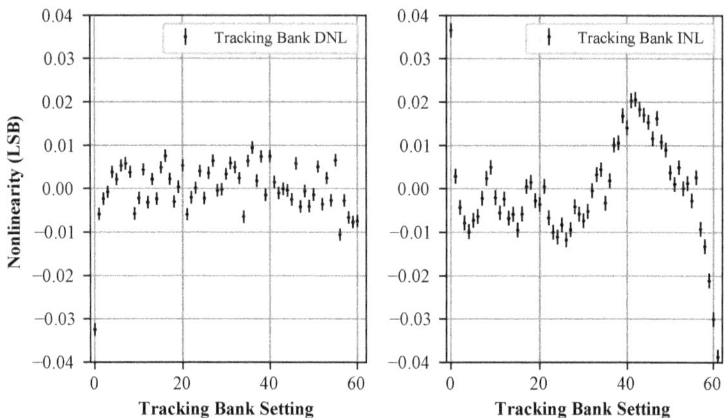

Figure 7. DCO tracking bank linearity measurements, including 95% confidence intervals. The absence of a dummy cell adjacent to cell 0 produces a small edge effect.

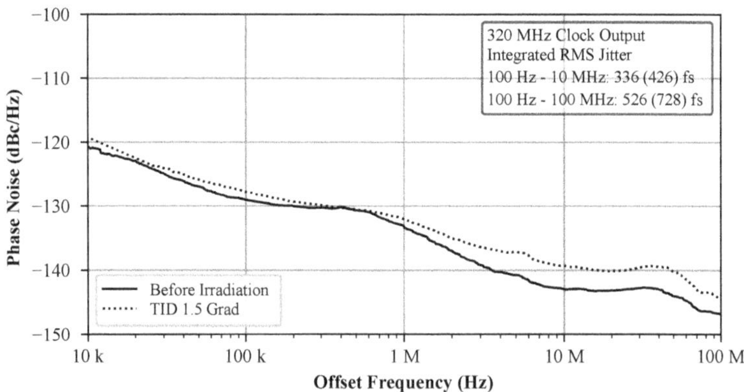

Figure 8. Closed-loop phase noise measurement of an All-Digital PLL circuit. Pre- and post-irradiation measurements are shown in solid and dashed lines, respectively. Values in parentheses indicate post-irradiation performance.

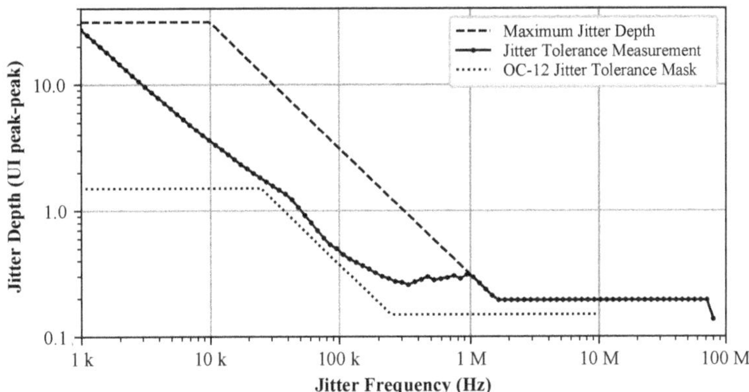

Figure 9. Closed-loop CDR jitter tolerance measurement obtained for a 320 Mbit s^{-1} input data rate. Jitter amplitude limits of the N4903B test setup are indicated with a dashed line.

5.3. Radiation Testing

For characterization of the circuit SEE sensitivity, irradiation with Heavy Ions was performed at the CRC Heavy Ion Facility in Louvain-la-Neuve, Belgium. Ions with Linear Energy Transfer (LET) between 3.3 MeV cm^2 mg^{-1} to 62.5 MeV cm^2 mg^{-1} were used for the measurement of the circuit cross-section. One of the PLL circuit clock outputs was instrumented with a transient phase measurement system, offering a resolution of 4 ps, as described in [30].

The experimentally determined SEE cross-section to heavy ion irradiation for two different phase excursion thresholds is shown in Figure 10. These measurements can be summarized as follows: A large cross-section but low magnitude sensitivity can be attributed to the spiral inductor. Temporary, positive frequency errors of the oscillator are stimulated by the irradiation of this large area [30]. This effect, while responsible for a saturation cross-section on the order of 1×10^{-3} cm^2 in this circuit, will not be discussed further here, as it exists in LC oscillators using planar inductors regardless of their loop architecture and also only results in small phase errors that scale with LET. As the underlying cause is a frequency error of the oscillator, the accumulated loop phase errors can be reduced by increasing the loop bandwidth. The responses of the circuit to this effect have been confined to phase excursions below 250 ps. This sensitivity has been previously discussed in [30].

The second class of SEE responses observed dominates the circuit cross-section for a 300 ps phase error threshold. An example of the observed transients for this class of effects is shown in Figure 11. The effect is characterized by a reduction in the DCO frequency for at least 20 µs. The origin of this sensitivity was identified in the bottom-pinning biasing scheme used for the PVT and acquisition banks (see Figure 4). While the turn-on time of switches M_{SW} and M_{PD} is in the order of one DCO oscillation period, establishing the bottom-pinning bias condition via M_{PIN} was found to require a long settling time, in line with the observed SEE response characteristics. For cells that are disabled during the irradiation, SETs affecting the enable input can temporarily turn on M_{SW} and M_{PD}, which disturbs the established bias condition by pulling V_A and V_B to the ground. While the biasing condition is re-established following this event, the DCO frequency remains slightly reduced, which results in an accumulation of phase errors in the loop. For the PVT cells, a stimulated frequency error of more than −50 ppm was measured during this settling time, and the sensitivity could be fully reproduced using SPICE simulations. A small number of these events at the highest experimental LET resulted in phase excursions up to 5 ns; however, no cycle slips of the PLL have been observed when operating with a 40 MHz reference clock frequency. Such high LET events are very unlikely to occur in high-energy physics experiments [31]. Nonetheless, this class of SEE can be mitigated using a different capacitor switch implementation, such as the one proposed in [32]. The large phase response to such a small frequency error is a result of its amplification by the BBPD-ADPLL architecture, which is inherently unable to provide a response proportional to the phase error which the long transient DCO frequency error causes. This might make PLL architectures with linear digital phase detectors or automatic loop filter gain adjustments when slewing is detected an attractive extension of this work in the future. No SEE responses could be identified that originated in the digital loop components (DLF, $\Sigma\Delta M$, FBDIV), highlighting the merits and effectiveness of the systematic SEE hardening methodology applied. To confirm this conclusion, the same digital PLL core was also tested with a radiation-tolerant ring DCO, and during its irradiation, no SEE responses exceeding 100 ps could be found.

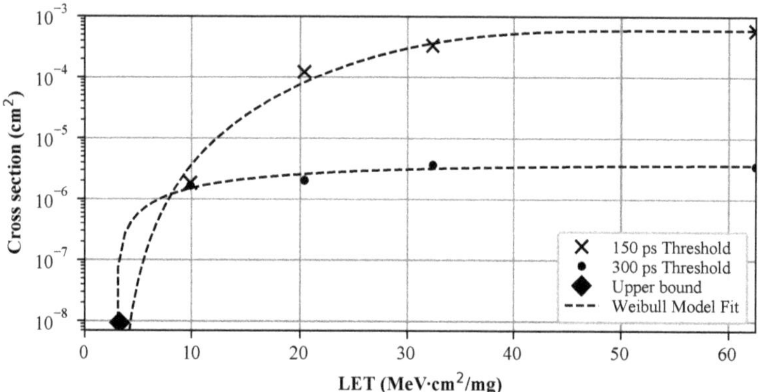

Figure 10. Experimental heavy ion cross-section for different phase excursion thresholds.

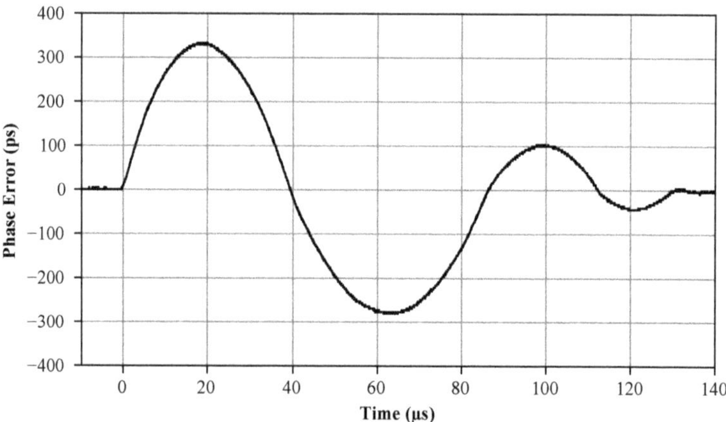

Figure 11. Example SEE phase transient stimulated by high LET ($62.5\,\mathrm{MeV\,cm^2\,mg^{-1}}$) ions. The parabolic phase response characteristic for Bang-Bang PLLs is observed. The loop transitions back to the random noise regime after 130 μs.

For the evaluation of its TID tolerance, the circuit was irradiated at the CERN X-ray irradiation facility at a temperature of −10 °C. This temperature is in the range of typical temperatures adopted for future detectors in the harsh radiation environments of the HL-LHC and is therefore expected to give a representative estimate of TID tolerance. A dose rate of 8.9 Mrad/h was used to accumulate a dose of 1.5 Grad over 170 h. The circuit was kept under bias permanently while repeatedly undergoing a characterization measurement routine.

During the irradiation, a consistent increase in the DCO free-running frequency was observed across all supply voltage and tuning control word settings. The evolution of the oscillation frequency at three different frequency settings is shown in Figure 12. The worst observed frequency shift remained within 0.8%, which is small compared to the total tuning range available (20%). This shift is a result of a change in common mode voltage and oscillation amplitude caused by the radiation-induced reduction in transconductance in the oscillator active devices [16]. Comparing this DCO design to the VCO design adopted in [9], we can find that the radiation-induced frequency shift is significantly reduced. As outlined in Section 4, the MOS varactors used in VCO designs tend to show a strong amplitude dependence, while the majority of capacitance in this DCO design is contributed to by MOM capacitances. The dominant nonlinear capacitance loading the tank in this design is the junction capacitance of the capacitor switches, which explains the significantly reduced sensitivity. This finding highlights the advantage of a varactorless DCO implementation over VCOs in terms of TID tolerance.

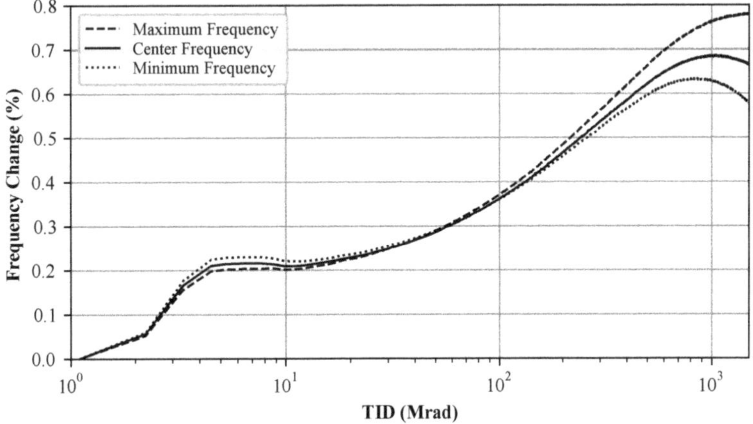

Figure 12. Evolution of DCO frequency under X-ray irradiation. A consistent frequency increase relative to the pre-irradiation value is seen for different frequency settings of the oscillator.

The circuit continued to operate normally up to 1.4 Grad. At this dose, a distinct drop of current consumption in the power domain supplying the DCO and divide-by-two prescaler was observed for the highest DCO frequency setting at the lower end of the specified supply voltage range. As the oscillation of the DCO did not cease and the PLL was still operational at this dose, this behavior can be explained by a failure of one of the TMR branches of the divide-by-two prescaler. Figure 6 shows the open loop phase noise performance of the oscillator at nominal operating conditions after the 1.5 Grad irradiation. Degradation is most pronounced at offset frequencies above 1 MHz and is dominated by the clock distribution network noise floor increase above 10 MHz. During the irradiation period, the circuit power dissipation was found to gradually reduce by about 15%.

Repeated measurements of the tracking bank tuning step size performed during the irradiation did not show a noticeable change. As the loop gain is otherwise determined primarily by the DLF, whose coefficients are insensitive to TID, the dynamic behavior of the PLL remained largely unaffected by the irradiation. This can be concluded from Figure 8, which compares the pre- and post-irradiation closed-loop noise performance. An increase in phase noise consistent with the previously discussed oscillator degradation can be observed. No locking failures have been observed during the test procedure, the PVT and acquisition banks were able to compensate for the oscillator frequency shift according to expectations and the output clock delay with respect to the reference clock remained repeatable across the irradiation. No failures of the DLF, $\Sigma\Delta M$ and divide-by-32 FBDIV were observed during the irradiation procedure, implying that the chosen synthesis strategy was adequate to achieve high radiation tolerance.

5.4. Summary

A comparison of this design to other radiation-tolerant PLL and CDR circuits reported in the literature is shown in Table 1. The proposed design outperforms the compared designs both in PLL FOM as well as in demonstrated TID tolerance. The remaining SEE sensitivity of the proposed circuit has been clearly identified, and a mitigation strategy has been proposed, which will allow eliminating any circuit cross-section for phase errors exceeding 250 ps, below which the inductor radiation effect will remain as the dominant contributor. In terms of area, the proposed circuit approaches the area of ring oscillator PLLs manufactured in the same technology [33], mainly because of the large area occupied by the loop filter capacitor in analog PLLs, which is realized as a digital integrator in this design.

Table 1. Comparison of radiation tolerant PLL/CDR designs.

Reference	This Work	[9]	[33]	[34]
Type	All-Digital PLL/CDR	Analog CDR	Analog CDR	Analog PLL
Oscillator	LC DCO	LC VCO	Ring VCO	LC VCO
Technology	65 nm	65 nm	65 nm	65 nm
Jitter (ps rms)	0.5	0.35	6.7	3.5
Power (mW)	11	34	7	18
Area (mm^2)	0.28	0.33	0.25	0.124
TID Tolerance (Mrad)	1500	350	600	250
FOM$_{PLL}$ (dB)	-235	-234	-215	-217

6. Conclusions

A radiation-tolerant All-Digital PLL and CDR circuit suitable for applications in high-energy physics was designed, implemented and tested. Different advantages of a varactorless DCO architecture combined with a digital loop implementation to replace sensitive analog charge pumps have been demonstrated in the context of radiation hardening. The successful demonstration of an implementation approach heavily utilizing automated TMR insertion, together with automated place and route methodologies, shows that design reuse and technology portability opportunities are retained even for circuits requiring radiation tolerance.

A crucial observation is that the remaining Single-Event Effect sensitivity has been identified in the DCO, which also remains the major mixed-signal component of the design. This underlines the continued difficulty of accurately predicting SEE sensitivities of such circuits and fully mitigating them during the design phase. This is in contrast to the digital design components, which were all successfully protected against SEU and SET by applying systematic and reliable design hardening techniques. A mitigation for the identified sensitivity has been proposed, which will allow pushing the envelope of radiation-tolerant All-Digital PLL circuits even further in future work.

Author Contributions: Conceptualization, S.B., S.K., P.L. and J.P.; Data curation, S.B.; Investigation, S.B.; Methodology, S.B., S.K. and J.P.; Project administration, P.M., P.L. and J.P.; Supervision, S.K., A.K., P.L. and J.P.; Visualization, S.B.; Writing—original draft, S.B.; Writing—review & editing, S.B., S.K., P.M., A.K., P.L. and J.P. All authors have read and agreed to the published version of the manuscript.

Funding: This work was funded by the Wolfgang Gentner Programme of the German Federal Ministry of Education and Research (grant no. 05E18CHA).

Acknowledgments: The authors would like to acknowledge Dipan Kar for their contribution to the LC DCO design.

Conflicts of Interest: The authors declare no conflict of interest. The funders had no role in the design of the study; in the collection, analyses, or interpretation of data; in the writing of the manuscript, or in the decision to publish the results.

References

1. ATLAS Collaboration. *Technical Design Report: A High-Granularity Timing Detector for the ATLAS Phase-II Upgrade*; Technical Report; CERN: Geneva, Switzerland, 2020.
2. *SpaceFibre—Very High-Speed Serial Link*; Standard; European Cooperation for Space Standardization: Noordwijk, NL, USA, 2019.
3. Orfanelli, S. The Phase 2 Upgrade of the CMS Inner Tracker. *Nucl. Instrum. Meth. A* **2020**, *980*, 164396. [CrossRef]
4. Mavis, D.; Eaton, P. Soft error rate mitigation techniques for modern microcircuits. In Proceedings of the 2002 IEEE International Reliability Physics Symposium. Proceedings. 40th Annual (Cat. No.02CH37320), Dallas, TX, USA, 7–11 April 2002; pp. 216–225. [CrossRef]
5. Staszewski, R.; Wallberg, J.; Rezeq, S.; Hung, C.M.; Eliezer, O.; Vemulapalli, S.; Fernando, C.; Maggio, K.; Staszewski, R.; Barton, N.; et al. All-digital PLL and transmitter for mobile phones. *IEEE J. Solid-State Circuits* **2005**, *40*, 2469–2482. [CrossRef]
6. Johnston, A. Radiation effects in advanced microelectronics technologies. *IEEE Trans. Nucl. Sci.* **1998**, *45*, 1339–1354. [CrossRef]
7. Prinzie, J.; Steyaert, M.; Leroux, P.; Christiansen, J.; Moreira, P. A single-event upset robust, 2.2 GHz to 3.2 GHz, 345 fs jitter PLL with triple-modular redundant phase detector in 65 nm CMOS. In Proceedings of the 2016 IEEE Asian Solid-State Circuits Conference (A-SSCC), Toyama, Japan, 7–9 November 2016; pp. 285–288. [CrossRef]
8. Loveless, T.D.; Massengill, L.W.; Holman, W.T.; Bhuva, B.L.; McMorrow, D.; Warner, J.H. A Generalized Linear Model for Single Event Transient Propagation in Phase-Locked Loops. *IEEE Trans. Nucl. Sci.* **2010**, *57*, 2933–2947. [CrossRef]
9. Biereigel, S.; Kulis, S.; Leitao, P.; Francisco, R.; Moreira, P.; Leroux, P.; Prinzie, J. A Low Noise Fault Tolerant Radiation Hardened 2.56 Gbps Clock-Data Recovery Circuit With High Speed Feed Forward Correction in 65 nm CMOS. *IEEE Trans. Circuits Syst. I Regul. Pap.* **2020**, *67*, 1438–1446. [CrossRef]
10. Jung, S.M.; Roveda, J.M. A radiation-hardened-by-design phase-locked loop using feedback voltage controlled oscillator. In Proceedings of the Sixteenth International Symposium on Quality Electronic Design, Santa Clara, CA, USA, 2–4 March 2015; pp. 103–106. [CrossRef]
11. Boulghassoul, Y.; Massengill, L.W.; Steinberg, A.L.; Bhuva, B.L.; Holman, W.T. Towards SET Mitigation in RF Digital PLLs: From Error Characterization to Radiation Hardening Considerations. *IEEE Trans. Nucl. Sci.* **2006**, *53*, 2047–2053. [CrossRef]
12. Loveless, T.D.; Massengill, L.W.; Bhuva, B.L.; Holman, W.T.; Witulski, A.F.; Boulghassoul, Y. A Hardened-by-Design Technique for RF Digital Phase-Locked Loops. *IEEE Trans. Nucl. Sci.* **2006**, *53*, 3432–3438. [CrossRef]
13. Loveless, T.D.; Massengill, L.W.; Bhuva, B.L.; Holman, W.T.; Reed, R.A.; McMorrow, D.; Melinger, J.S.; Jenkins, P. A Single-Event-Hardened Phase-Locked Loop Fabricated in 130 nm CMOS. *IEEE Trans. Nucl. Sci.* **2007**, *54*, 2012–2020. [CrossRef]
14. Gao, Y.; Lou, J.; Zhang, J.; Li, L.; Xu, F. Single-event transients in LC-tank VCO. In Proceedings of the 2012 International Workshop on Microwave and Millimeter Wave Circuits and System Technology, Chengdu, China, 19–20 April 2012; pp. 1–4. [CrossRef]
15. Zhang, Z.; Chen, L.; Djahanshahi, H. A Hardened-By-Design Technique for LC-Tank Voltage Controlled Oscillator. In Proceedings of the 2018 IEEE Canadian Conference on Electrical Computer Engineering (CCECE), Quebec, QC, Canada, 13–16 May 2018; pp. 1–4. [CrossRef]
16. Prinzie, J.; Christiansen, J.; Moreira, P.; Steyaert, M.; Leroux, P. Comparison of a 65 nm CMOS Ring- and LC-Oscillator Based PLL in Terms of TID and SEU Sensitivity. *IEEE Trans. Nucl. Sci.* **2017**, *64*, 245–252. [CrossRef]
17. Chen, Y.P.; Massengill, L.W.; Bhuva, B.L.; Holman, W.T.; Loveless, T.D.; Robinson, W.H.; Gaspard, N.J.; Witulski, A.F. Single-Event Characterization of Bang-bang All-digital Phase-locked Loops (ADPLLs). *IEEE Trans. Nucl. Sci.* **2015**, *62*, 2650–2656. [CrossRef]
18. Chen, Y.P.; Massengill, L.W.; Kauppila, J.S.; Bhuva, B.L.; Holman, W.T.; Loveless, T.D. Single-Event Upset Characterization of Common First- and Second-Order All-Digital Phase-Locked Loops. *IEEE Trans. Nucl. Sci.* **2017**, *64*, 2144–2151. [CrossRef]
19. Chen, Y.P.; Massengill, L.W.; Sternberg, A.L.; Zhang, E.X.; Kauppila, J.S.; Yao, M.; Amort, A.L.; Bhuva, B.L.; Holman, W.T.; Loveless, T.D. Time-Domain Modeling of All-Digital PLLs to Single-Event Upset Perturbations. *IEEE Trans. Nucl. Sci.* **2018**, *65*, 311–317. [CrossRef]
20. Alexander, J. Clock recovery from random binary signals. *Electron. Lett.* **1975**, *11*, 541–542. [CrossRef]
21. Staszewski, R.; Hung, C.M.; Barton, N.; Lee, M.C.; Leipold, D. A digitally controlled oscillator in a 90 nm digital CMOS process for mobile phones. *IEEE J. Solid-State Circuits* **2005**, *40*, 2203–2211. [CrossRef]
22. Xu, Z.; Miyahara, M.; Okada, K.; Matsuzawa, A. A 3.6 GHz Low-Noise Fractional-N Digital PLL Using SAR-ADC-Based TDC. *IEEE J. Solid-State Circuits* **2016**, *51*, 2345–2356. [CrossRef]
23. Kuo, F.W.; Babaie, M.; Chen, H.N.R.; Cho, L.C.; Jou, C.P.; Chen, M.; Staszewski, R.B. An All-Digital PLL for Cellular Mobile Phones in 28-nm CMOS with -55 dBc Fractional and -91 dBc Reference Spurs. *IEEE Trans. Circuits Syst. I Regul. Pap.* **2018**, *65*, 3756–3768. [CrossRef]

24. Bonfanti, A.; Levantino, S.; Samori, C.; Lacaita, A. A varactor configuration minimizing the amplitude-to-phase noise conversion in VCOs. *IEEE Trans. Circuits Syst. I Regul. Pap.* **2006**, *53*, 481–488. [CrossRef]
25. Bonacini, S.; Valerio, P.; Avramidou, R.; Ballabriga, R.; Faccio, F.; Kloukinas, K.; Marchioro, A. Characterization of a commercial 65 nm CMOS technology for SLHC applications. *J. Instrum.* **2012**, *7*, P01015. [CrossRef]
26. Hershberg, B.; Raczkowski, K.; Vaesen, K.; Craninckx, J. A 9.1–12.7 GHz VCO in 28nm CMOS with a bottom-pinning bias technique for digital varactor stress reduction. In Proceedings of the ESSCIRC 2014—40th European Solid State Circuits Conference (ESSCIRC), Venice Lido, Italy, 22–26 September 2014; pp. 83–86. [CrossRef]
27. Nikolaou, A.; Bucher, M.; Makris, N.; Papadopoulou, A.; Chevas, L.; Borghello, G.; Koch, H.D.; Faccio, F. Modeling of High Total Ionizing Dose (TID) Effects for Enclosed Layout Transistors in 65 nm Bulk CMOS. In Proceedings of the 2018 International Semiconductor Conference (CAS), Sinaia, Romania, 10–12 October 2018; pp. 133–136. [CrossRef]
28. Kulis, S. Single Event Effects mitigation with TMRG tool. *J. Instrum.* **2017**, *12*, C01082. [CrossRef]
29. Jara Casas, L.M. *DRAD Results Obtained during Irradiation Campaigns*; Technical Report; CERN: Geneva, Switzerland, 2020.
30. Biereigel, S.; Kulis, S.; Leroux, P.; Moreira, P.; Kölpin, A.; Prinzie, J. Single-Event Effect Responses of Integrated Planar Inductors in 65 nm CMOS. *IEEE Trans. Nucl. Sci.* **2021**. [CrossRef]
31. Huhtinen, M.; Faccio, F. Computational method to estimate single event upset rates in an accelerator environment. *Nucl. Instrum. Meth. A* **2000**, *450*, 155–172. [CrossRef]
32. Sjoland, H. Improved switched tuning of differential CMOS VCOs. *IEEE Trans. Circuits Syst. II Analog. Digit. Signal Process.* **2002**, *49*, 352–355. [CrossRef]
33. Moustakas, K.; Rymaszewski, P.; Hemperek, T.; Krüger, H.; Vogt, M.; Wang, T.; Wermes, N. A Clock and Data Recovery Circuit for the ALTAS/CMS HL-LHC Pixel Front End Chip in 65 nm CMOS Technology. In Proceedings of the Topical Workshop on Electronics for Particle Physics TWEPP2019, Santiago de Compostela, Spain, 2–6 September 2019; Volume 2, p. 6. [CrossRef]
34. Mazza, G.; Panati, S. A Compact, Low Jitter, CMOS 65 nm 4.8–6 GHz Phase-Locked Loop for Applications in HEP Experiments Front-End Electronics. *IEEE Trans. Nucl. Sci.* **2018**, *65*, 1212–1217. [CrossRef]

Article

Novel Full TMR Placement Techniques for High-Speed Radiation Tolerant Digital Integrated Circuits

Karel Appels * and Jeffrey Prinzie

Deptartment of Electrical Engineering (ESAT), KU Leuven, 3000 Leuven, Belgium; jeffrey.prinzie@kuleuven.be
* Correspondence: karel.appels@kuleuven.be; Tel.: +32-477-625717

Received: 21 October 2020; Accepted: 12 November 2020; Published: 17 November 2020

Abstract: This paper presents a novel physical implementation methodology for high-speed Triple Modular Redundant (TMR) digital integrated circuits for harsh radiation environment applications. An improved distributed approach is presented to constrain redundant branches of Triple Modular Redundant (TMR) digital logic cells using repetitive, interleaved micro-floorplans. To optimally constrain the placement of both sequential and combinational cells, the TMR netlist is used to segment the the logic into unrelated groups allowing sharing without compromising reliability. The technique was evaluated in a 65 nm bulk CMOS technology and a comparison is made to conventional methods.

Keywords: triple modular redundancy; 65 nm CMOS technology; single event effects; radiation hardening by design; digital integrated circuits

---

1. Introduction

Single Event Effect (SEEs) are undesired erroneous effects in digital integrated circuits caused by ionizing radiation. With CMOS device scaling, SEEs have become an increasingly important reliability concern leading to severe soft-error rates in advanced systems [1]. Not only in nuclear instrumentation or space applications, even in critical commercial applications such as autonomous transport systems soft errors have become an increasingly important concern. With the growing complexity of digital circuits and clock frequencies, the overhead of redundancy should be reduced to a minimum. This paper addresses both Single Event Transients (SETs) and Single Event Upsets (SEUs). SETs are temporal erroneous signals which originate from the charges generated by the incident particles which are collected by the transistor in a combinational cell. They will recover over time, as can be seen on Figure 1 [2]. However if the SET propagates to a sequential cell like a flip-flop and occurs within the setup and hold times of the registers near a clock edge, the SET is latched leading to an incorrect logical state which is also known as an SEU as is shown in Figure 1. The probability of such latching increases proportionally with higher clock frequencies [3,4]. SEUs can also occur when charged particles directly hit sequential digital circuits such as latches and flip-flops. When the register involves a bit-flip, this erroneous signal may remain in the digital system and can even propagate to other digital modules resulting in a failure. For example, an SEU could change the state of an Finite State Machine (FSM) temporarily impacting the entire system. SEUs can thus originate from direct upsets in the registers or as a result from SETs in the combinational logic, latched during clocking.

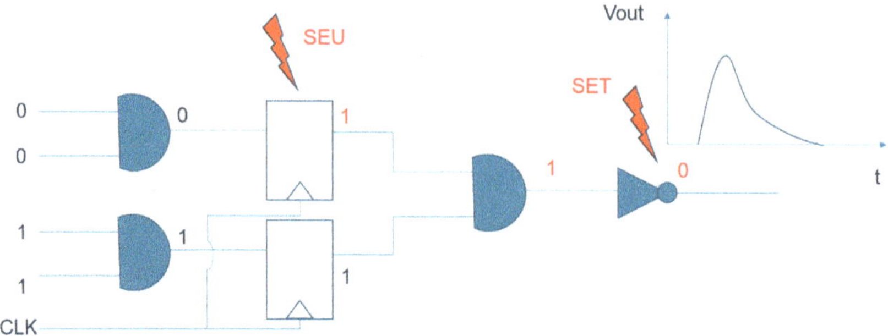

Figure 1. Single Event Transients (SETs) and Single Event Upsets (SEUs) in a digital circuit.

Triple Modular Redundancy (TMR) can be used to protect digital logic from SEEs. It uses redundant logic with majority voters to correct logic signals [5]. TMR only works if only single errors occur in the digital logic, hence Multi Bit Upsets (MBUs) in common logic signals can be catastrophic for TMR. Nowadays complementary metal oxide semiconductor (CMOS) technologies have scaled to the point that MBUs have become a serious concern since a single particle can affect multiple gates simultaneously [6–8]. This was less important in old CMOS technologies where single particles only affected single digital cells. However, with proper placement techniques, the fault tolerance can be ensured without compromising speed or power consumption in the design which is addressed in this paper.

Historically, several methods were developed to address trade-offs in TMR designs like power consumption and area efficiency. Full TMR is the most robust and most complete form of redundancy. In this approach, both the flip-flops, clock-tree and combinational logic cells are triplicated [9,10]. However, the drawbacks of this approach are the high number of resources (digital gates) and power consumption. Nevertheless it is the most solid and secure form of TMR. One of the competing methods is temporal time redundancy [11]. In this method, only flip-flops are triplicated which are clocked with 3 skewed clocks. The combinational logic is not redundant. The skew between the clocks must be larger than any possible SET, hence only one flip-flop could possibly latch an SET. This method has proven its usage in many applications [12]. However its major drawback is its limited clock frequency since clock skew places strict timing constraints on the design typically resulting in sub-GHz timing performance. Henceforth, many high-speed mixed-signal digital modules are based the original TMR approach. Additionally, Error Detection and Correction (EDAC) codes can also be used for radiation hardness assurance. They are usually placed surrounding the sequential cells to correct for any SEUs. Depending on the coding scheme, EDACs are also vulnerable to MBUs which must be mitigated as well and might be more difficult to ensure compared to TMR. The coding and decoding logic also adds to additional timing overhead which might significantly slow down the critical datapaths.

As indicated above, TMR only works if single errors occur. In deep-submicron technologies, proper physical implementation is required to ensure no MBUs occur between cells of the same TMR logic branch. This paper presents an innovative optimised physical placement methodology for full TMR design.

2. 3 Block Approach

One the most frequently used methods for physical implementation of TMR designs is by separating A-B-C logic in 3 different areas. Hence, a floor plan is created with 3 blocks (named A, B and C) as is shown on Figure 2. All flipflops are constrained to their specified block A-C, the combinational logic and the clock tree will intrinsically follow the placement of the flipflops. However, each sequential net has 6 cross domain (A-C) interconnections(voter inputs) leading to long nets that transverse across the entire floorplan. Hence, the power consumption increases, and the net connections between the flipflops become congested.

As the design size increases, the length and the routing will be more complex resulting in increased routing congestion and power consumption due to additional buffers inserted by the place-and-route tools in the cross domain voted nets to meet timing constraints. Hence, power consumption and routing complexity is the main concern limiting its usability small or low frequency designs.

3. Advanced Placement Methods

In this section, an optimal physical implementation scheme is proposed: an interleaved method [13] and an improved interleaved method, overcoming power and area trade-offs.

3.1. Interleaved Floorplan

One of the limitation of the 3 block floorplan is the limited freedom of place-and-route tools to place critical logic relatively close together (while respecting minimal spacing). The interleaved approach, presented in Reference [13], uses a semi-distributed placement method to ensure maximal freedom to the place-and-route tools to optimize the design. It is based on the conventional 3 block approach but uses an interleaved placement constraining method of many small A,B,C sections, shown on Figure 2.

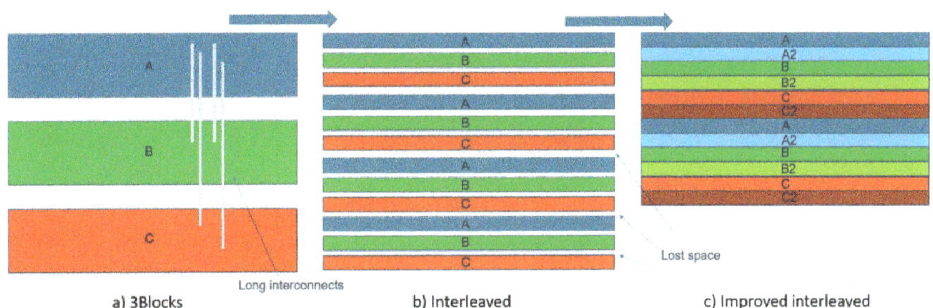

Figure 2. Different placement methods. (**a**) 3Blocks (**b**) Interleaved (**c**) Improved interleaved.

Instead of 3 large blocks, there are multiple repeating small regions, allowing cells of A-C branches to be placed at different vertical spots. Each region has the same fixed height. The distance between each region ensures that Multi Bit Upsets (MBUs) cannot occur. As the design size expands, the height of the regions does not expands but only the number of vertical regions increases. Consequently, vertical connections between voters always cross the same narrow placement region and have equal lengths, regardless of the design size. Therefore, the place and route tool has much more freedom to place the cells vertically and much closer to each other. This is a significant improvement compared to the 3 block implementation where voter connections have to cross a significant portion of the design. To constrain the flip-flops and the data path cells, a trace-back algorithm was used to find the corresponding logic tree from a source sequential element [13].

3.2. Improved Interleaved Method

The main drawback of the interleaved method is the lost space between the placement sections to ensure proper spacing between TMR branches. To improve the interleaved method, the same principle can be applied, however the lost space can be recovered by filling its empty area with unrelated digital logic. Again, this method allows a semi-distributed placement to create maximal freedom to place-and-route tools to optimise the design as shown in Figure 2. In the proposed method, a floorplan is made using 6 physical constrain groups (A1, A2, B1, B2, C1, C2), or denoted ABC1 and ABC2. Each group has a height equal to or larger than the required spacing distance to prevent MBUs and occupies the entire width of the design. Vertically, all groups are repetitive to fill the vertical design space (e.g., A1-A2-B1-B2-C1-C2-A1-A2-etc.). A TMR logic branch is placed in either ABC1 or ABC2. As such, one group acts as spacer to the other and is allowed to share upsets if the cells do not have a common datapath. To ensure maximal area efficiency, TMR paths are balanced between ABC1 and ABC2 if they do not share a common combinational path and thus are allowed to share multi-cell upsets.

The advantage of this approach is the elimination of lost space, as can be seen on Figure 2. To balance the logic between ABC1 and ABC2, a segmentation algorithm is used to detect if a logic tree is connected to an existing tree in ABC1 or ABC2. If one of the cells in a logic tree is already in ABC1 or ABC2, then this entire instance group is placed within that group. If this is not the case, the cells are placed in the least filled group. The total area of each group is continuously maintained in order to balance them in terms of area. The segmentation algorithm is shown graphically in Figure 3. The time needed for the segmentation algorithm is negligible compared to the duration of the place and route tools. Furthermore the time needed to backtrace all combinational cells from netlist to the 3 different A-B-C branches is substantially longer than the segmentation algorithm itself, although also negligible compared to the place and route tools.

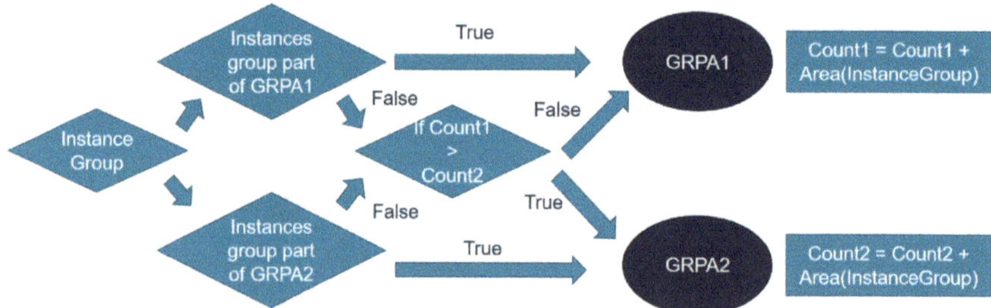

Figure 3. Segmentation algorithm: Triple Modular Redundancy (TMR) datapaths are balanced between 2 Groups representing the floorplan groups ABC1 and ABC2.

4. Simulated Analysis

Different comparative studies were performed with either the interleaved, the improved interleaved method or the standard 3 block approach to evaluate the efficiency and performance of this new placement and floorplanning technique. As benchmark, a design with eight identical and independent high-speed counters was used. To introduce more complicated standardised data paths, the counter dimensions (widths) and count of the benchmark models were varied. Larger counter widths would indicate a more complex datpath. More counters resemble a larger overall design in order to evaluate if the techniques scale with increasing design widths. The designs have been implemented and analyzed using the Cadence Innovus Computer Aided Design (CAD) tools. To guarantee a timing critical design, the timing limitations were selected to be near the technology boundaries. In the analysis, power consumption, net length, net capacitance and routing density were evaluated for each method. The slice height and spacing of the interleaved techniques was chosen as 7.2 μm, whereas the 3 block technique has 7.2 μm block spacing. This number aligned with the cells' row heights. Finally, the timing, power and region reports from pace-and-route tools were extracted.

Figure 4 shows the routed designs of 8 × 16 bit counters for the 3 different methods. We did not observe a considerable difference in performance variation for both the interleaved techniques for varying counter dimensions. Compared to the standard 3 block strategy, it is evident that the suggested interleaved and improved interleaved methods result in considerably reduced complexity. In particular, the vertical routing difficulty reduces significantly since the place-and-route tool has more freedom to efficiently place the standard cells closer to each other. Comparing the interleaved and the improved interleaved method, it is clear that the difference between the routing complexity is relatively small, as expected. However the area efficiency (standard cell density) to place the design is improved. The Amouba view of the design is shown on Figure 5. Each colour represents a triplicated counter. In the 3 block method, the counters are distributed across the 3 bulky blocks. In the interleaved method there is a much better grouping of the counters compared to the 3 block method since the cells can be spaced more closely. Again, the improved interleaved method demonstrates its advantage in the fact that independent combinational path acts as a spacing distance for the other, ensuring maximal area efficiency.

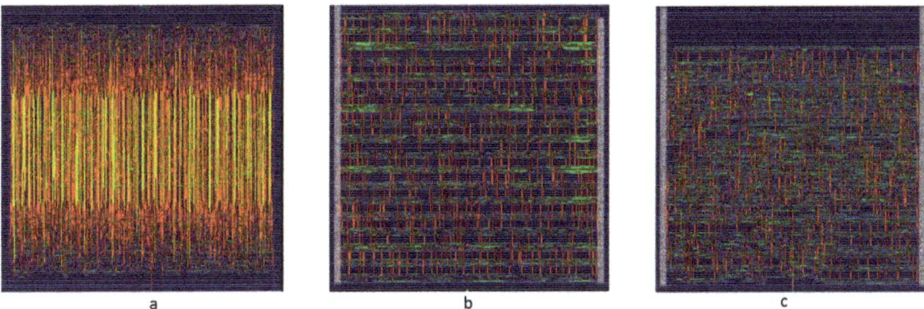

Figure 4. Routing of the 3 different implementation strategies: (**a**) 3Blocks (**b**) Interleaved (**c**) Improved interleaved.

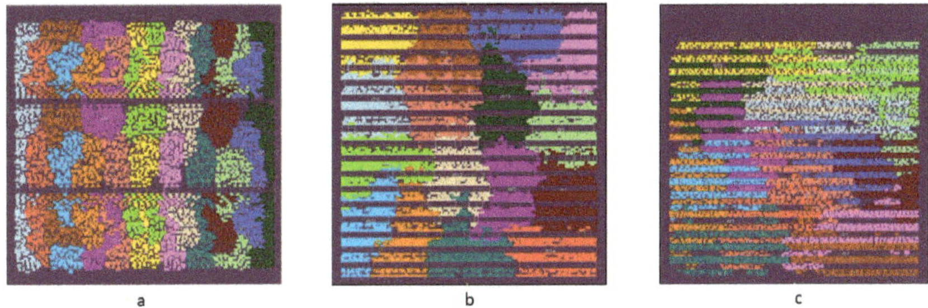

Figure 5. Amouba view of the investigated design. Each triplicated counter has a different color to highlight the placement strategy. (**a**) 3Blocks (**b**) Interleaved (**c**) Improved Interleaved.

The distances between the cells of same TMR branches are shown on Figure 6. The Distances between the cells of the 3 block implementations are substantially larger than those within both the interleaved methods. Most cells are spaced within a range of 45 μm, which corresponds to approximately 3 elementary, vertical interleaved banks. It is evident from this consequence that the placement engine has more liberty to put cells closer without compromising the radiation hardness. The average cell distance between

A-branch and B-or C-cells for the 3 block technique shows two peaks corresponding with the A-B-C and the A-C distance.

A comparison of the net lengths distribution is shown on Figure 7. The histogram demonstrates that due to voting interconnections, a significant part of the nets for the 3 block method has 1/3 to 1/2 of the design size. This peak is no longer present in the suggested interleaved techniques. In this case, most networks that are interconnected have a net length of 25 µm or less, though this is still design specific. However, in the interleaved model there are still a few long nets. By analysing pre-Clock Tree Synthesis (CTS) and post-CTS histograms, it becomes apparent that these longer connections originate from the clock tree.

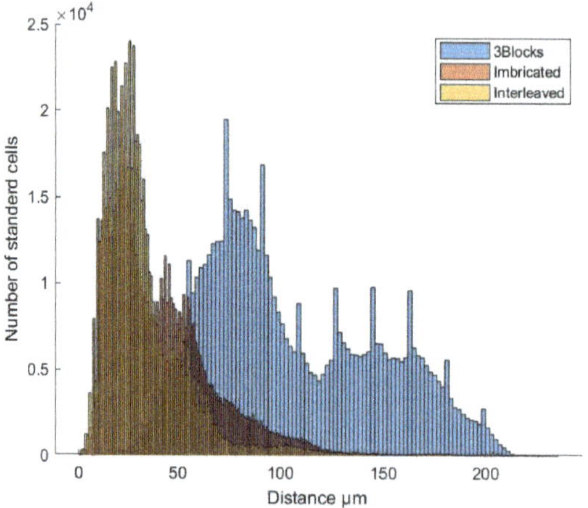

Figure 6. TMR branch logic spacing for the 8 × 16 bit design post-Clock Tree Synthesis (CTS).

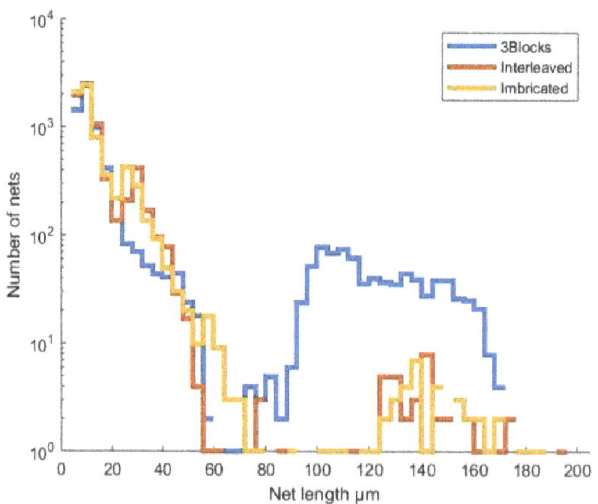

Figure 7. Net length histogram for the 8 × 16 bit design post-PCTS.

Clock trees A-C in the interleaved implementations are now distributed throughout the entire design, while the clock tree was only locally placed in each of the 3 regions in the 3 block approach. This is shown on Figure 8. Each colour represent one of the clock branches.

The average metal density of the different metal layers is shown on Figure 9. M1 and all the horizontal layers show no compelling difference since these layers are not used to to interconnect cells. On the other hand, M2 and M4 shows a significant difference due to routing between voter cells. The density in vertical routing layer M2 is almost reduced by half compared to the 3 block method and there is almost no metal density in layer 4, indicating the significance of the proposed strategies. In metal layer 3 there is a slight increase in density of the proposed method. When comparing the interleaved with the proposed improved implementation, a higher gate density is achieved which, as a result, also increases the local corresponding routing density, mainly in M3. Additionally, In the 3 blocks method the placement blocks are much higher which results in a more favourable vertical distribution. In the interleaved methods there are many more shorter blocks which results in a different placement vertically. However, the main figure is the reduction in extremely long M4 nets across the floorplan.

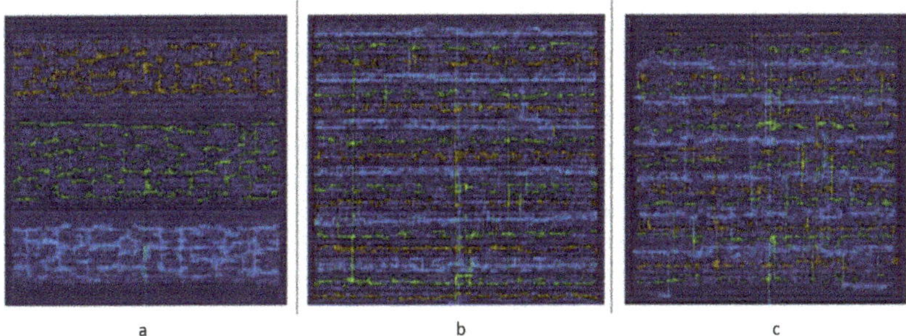

Figure 8. Clock tree coloured for the three branches A, B and C: (**a**) Clocktree 3Blocks method (**b**) Clocktree interleaved method (**c**) Clocktree interleaved 2.0 method.

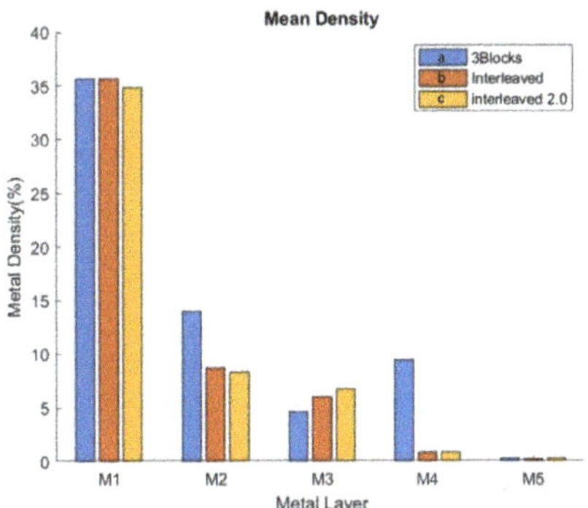

Figure 9. Average design metal density for different metal layers: (**a**) 3Block (**b**) Interleaved (**c**) Improved interleaved.

A comparison of the power consumption of the 8×16 bit counter is shown in Table 1. These numbers were extracted after routing and CTS. The internal power is the power consumption of the unloaded standard cells, switching power is the dynamic power consumption due to the switching of the capacitive loads (cells and nets) and the Total capacitance is the sum of all net and input capacitances of the cells. As can be expected, the internal power does not change significantly since the design remains almost identical and the only difference between all methods arises from different buffers which is only a small fraction of the total internal power consumption. The total net lengths of both interleaved methods are significantly smaller compared to the 3 block approach due to the optimal placement by the place-and-route tools. As a result, the total capacitance of the design reduces proportionally as is the dynamic power consumption. Since the main reduction is a result from avoiding long voters interconnects, the improvements become more significant as the design size increases. Therefore, this technique scales well with larger digital designs.

Table 1. Power consumption table.

	3Blocks	Interleaved	Improved Interleaved	Diff 3Block-Interleaved (%)	Diff 3Block-Improved Interleaved (%)
Internal power (mW)	18.7	18.6	18.6	−0.3	−0.3
Switching power (mW)	11.1	6.8	6.7	−38.3	−39.5
Total power (mW)	29.8	25.5	25.3	−14.5	−14.9
Total Capacitance (pF)	36.3	24.4	24.3	−32.7	−33

Finally, these results were extracted by evaluating a 65 nm CMOS technology. This methodology however scales well to smaller CMOS nodes. Firstly, in smaller nodes, designs often become more complex and the need for optimal placement increases significantly. Secondly, smaller nodes become more susceptible to SEEs meaning the proposed methods will be increasingly mandatory. Finally, since routing becomes the strongest contributor to power consumption in deep submicron technologies, the proposed methods will show an increasing improvement as devices scale down.

5. Conclusions

The major advantage of this distributed placement approach is that place-and-route tools have more freedom to distribute logic across the floorplan. In contrast to the 3 block approach, interconnections between voters do not need to cross a large center block that results in major routing complexity and power consumption. The total net length is drastically reduced since the connected logic can be placed more closely together, still ensuring minimal spacing for SEEs. As a consequence, the switching power is reduced. With the proposed improved distributed method, by using the placement balancing between ABC1 and ABC2 and using one group as MBU spacer for the other, the area efficiency is maximized compared to the earlier reported interleaved placement strategy.

Author Contributions: Conceptualisation, K.A. and J.P.; methodology, K.A. and J.P.; validation, K.A. and J.P.; investigation, K.A. and J.P.; writing—original draft preparation, K.A. and J.P. All authors have read and agreed to the published version of the manuscript.

Funding: This research was funded by FWO.

Conflicts of Interest: The authors declare no conflict of interest. The founding sponsors had no role in the design of the study; in the collection, analyses, or interpretation of data; in the writing of the manuscript, and in the decision to publish the results.

Electronics **2020**, *9*, 1936

References

1. Nsengiyumva, P.; Ball, D.R.; Kauppila, J.S.; Tam, N.; McCurdy, M.; Holman, W.T.; Alles, M.L.; Bhuva, B.L.; Massengill, L.W. A Comparison of the SEU Response of Planar and FinFET D Flip-Flops at Advanced Technology Nodes. *IEEE Trans. Nucl. Sci.* **2016**, *63*, 266–272. [CrossRef]
2. Chen, C.-H.; Knag, P.; Zhang, Z. Characterization of Heavy-Ion-Induced Single-Event Effects in 65 Nm Bulk CMOS ASIC Test Chips. *IEEE Trans. Nucl. Sci.* **2014**, *61*, 2694–2701. [CrossRef]
3. Benedetto, J.; Eaton, P.; Avery, K.; Mavis, D.; Gadlage, M.; Turflinger, T.; Dodd, P.E.; Vizkelethyd, G. Heavy ion-induced digital single-event transients in deep submicron Processes. *IEEE Trans. Nucl. Sci.* **2004**, *51*, 3480–3485. [CrossRef]
4. Benedetto, J.M.; Eaton, P.H.; Mavis, D.G.; Gadlage, M.; Turflinger, T. Digital Single Event Transient Trends with Technology Node Scaling. *IEEE Trans. Nucl. Sci.* **2006**, *53*, 3462–3465. [CrossRef]
5. Ruano, O.; Reviriego, P.; Maestro, J.A. Automatic insertion of selective TMR for SEU mitigation. In Proceedings of the RADECS, Gothenburg, Sweden, 16–21 September 2018; pp. 284–287.
6. Evans, A.; Glorieux, M.; Alexandrescu, D.; Polo, C.B.; Ferlet-Cavrois, V. Single event multiple transient (SEMT) measurements in 65 nm bulk technology. In Proceedings of the 16th European Conference on Radiation and Its Effects on Components and Systems (RADECS), Bremen, Germany, 19–23 September 2016. [CrossRef]
7. Zhang, H.; Jiang, H.; Assis, T.R.; Ball, D.R.; Narasimham, B.; Anvar, A.; Massengill, L.W.; Bhuva, B.L. Angular Effects of Heavy-Ion Strikes on Single-Event Upset Response of Flip-Flop Designs in 16-nm Bulk FinFET Technology. *IEEE Trans. Nucl. Sci.* **2017**, *64*, 491–496. [CrossRef]
8. Bhuva, B.L.; Tam, N.; Massengill, L.W.; Ball, D.; Chatterjee, I.; McCurdy, M.; Alles, M.L. Multi-Cell Soft Errors at Advanced Technology Nodes. *IEEE Trans. Nucl. Sci.* **2015**, *62*, 2585–2591. [CrossRef]
9. Balasubramanian, P.; Prasad, K. A Fault Tolerance Improved Majority Voter for TMR System Architecture. *WSEAS Trans. Circ. Syst.* **2016**, *15*, 108–122.
10. Shinohara, K.; Watanabe, M. A double or triple module redundancy model exploiting dynamic reconfigurations. In Proceedings of the NASA/ESA Conference on Adaptive Hardware and Systems, Noordwijk, The Netherlands, 22–25 June 2008; pp. 114–112.
11. Hasanbegovic, A.; Aunet, S. Heavy Ion Characterization of Temporal-, Dual- and Triple Redundant Flip-Flops Across a Wide Supply Voltage Range in a 65 nm Bulk CMOS Process. *IEEE Trans. Nucl. Sci.* **2016**, *63*, 2962–2970. [CrossRef]
12. Schmidt, R.; Garcia-Ortiz, A.; Fey, G. Temporal redundancy latch-based architecture for soft error mitigation. In Proceedings of the IEEE 23rd International Symposium on On-Line Testing and Robust System Design (IOLTS), Thessaloniki, Greece, 3–5 July 2017; pp. 240–243.
13. Prinzie, J.; Appels, K.; Szymon, K. Optimal Physical Implementation of Radiation Tolerant High-Speed Digital Integrated Circuits in Deep-Submicron Technologies. *Electronics* **2019**, *8*, 4.

Publisher's Note: MDPI stays neutral with regard to jurisdictional claims in published maps and institutional affiliations.

Article

A High-Reliability Redundancy Scheme for Design of Radiation-Tolerant Half-Duty Limited DC-DC Converters

Solomon Mamo Banteywalu [1], Getachew Bekele [1], Baseem Khan [2], Valentijn De Smedt [3] and Paul Leroux [3,*]

[1] Department of Electrical and Computer Engineering, Addis Ababa Institute of Technology, Addis Ababa University, Addis Ababa 1176, Ethiopia; solomonmamo.banteywalu@student.kuleuven.be (S.M.B.); getachewb@gmail.com (G.B.)
[2] Department of Electrical and Computer Engineering, Institute of Technology, Hawassa University, Hawassa 05, Ethiopia; baseem.khan04@gmail.com
[3] Department of Electrical Engineering, Faculty of Engineering Technology, Katholieke Universiteit Leuven, 3000 Leuven, Belgium; valentijn.desmedt@kuleuven.be
* Correspondence: paul.leroux@kuleuven.be

Abstract: Redundancy techniques are commonly used to design radiation- and fault-tolerant circuits for space applications, to ensure high reliability. However, higher reliability often comes at a cost of increased usage of hardware resources. Triple Modular Redundancy (TMR) ensures full single fault masking, with a >200% power and area overhead cost. TMR/Simplex ensures full single fault masking with a slightly more complicated circuitry, inefficient use of resource and a >200% power and area overhead cost, but with higher reliability than that of TMR. In this work, a high-reliability Spatial and Time Redundancy (TR) hybrid technique, which does not abandon a working module and is applicable for radiation hardening of half-duty limited DC-DC converters, is proposed and applied to the design of a radiation-tolerant digital controller for a Dual-Switch Forward Converter. The technique has the potential of double fault masking with a <2% increase in resource overhead cost compared to TMR. Moreover, for a Simplex module failure rate, λ, of 5%, the Reliability Improvement Factor (RIF) over the Simplex system is 20.8 and 500 for the proposed technique's two- and three-module implementations, respectively, compared to a RIF over the Simplex system of only 7.25 for TMR and 14.3 for the regular TMR/Simplex scheme.

Keywords: triple modular redundancy TMR; time redundancy (TR); TMR/Simplex; reliability improvement factor (RIF); half-duty limited DC-DC converter

Citation: Banteywalu, S.M.; Bekele, G.; Khan, B.; De Smedt, V.; Leroux, P. A High-Reliability Redundancy Scheme for Design of Radiation-Tolerant Half-Duty Limited DC-DC Converters. *Electronics* **2021**, *10*, 1146. https://doi.org/10.3390/electronics10101146

Academic Editor: Gianluca Traversi

Received: 14 April 2021
Accepted: 10 May 2021
Published: 12 May 2021

Publisher's Note: MDPI stays neutral with regard to jurisdictional claims in published maps and institutional affiliations.

1. Introduction

Space abounds with radiation sources that challenge the normal and stable operations of electronic circuits on board spacecraft and satellites. Radiation may cause circuit malfunction or, in the worst case, complete failure. This necessitates the radiation-tolerant design of electronic circuits intended to be used in the space environment.

To ensure radiation tolerance, different methods have been proposed and are actually being used in space [1–12]. Using radiation-hardened devices or fault-tolerant designs are the most common methods. Redundancy is one of the solutions applied to ensure a circuit is able to tolerate faults induced by radiation.

Redundancy techniques are used to construct electronic circuits that can endure radiation effects in a space environment, and can operate reliably in the presence of radiation-induced faults occurring in hardware and software. The use of SRAM FPGAs for the design of digital circuits for space applications has increased recently due to the advantages these technologies provide, which include flexibility in terms of quick turn-around time and on-orbit reconfiguration capability. Considerable work has been undertaken for the use of redundancy techniques for space applications implemented on SRAM-based FPGAs [1–4,13–20].

N Modular Redundancy (NMR) uses N copies of a module, where N is usually odd, with a voting system to tolerate faults in up to $(N - 1)/2$ modules. Triple Modular Redundancy (TMR) is currently widely used to mitigate radiation-induced faults and is considered to have "saved" several space missions [1–12].

TMR offers significantly better reliability than Simplex (unmitigated system) for short mission times. It is mostly used in applications in which mission times are typically short compared to component life. This is because, after the first failure, TMR is equivalent to a system of two modules in series, with the failure rate double that of a Simplex system. If Rm is the reliability of one of the modules (Simplex system), the reliability equation of the TMR system if an ideal voter is assumed is given by [21]:

$$R_{TMR} = 3R_m^2 - 2R_m^3 \tag{1}$$

A special form of TMR that combines advantages of TMR and Simplex in one system is TMR/Simplex, which is a reconfigurable, masking redundancy method in which differences in the outputs of the modules are detected and cause a reconfiguration of the TMR system. In particular, it detects a single module failure; the failed module and one of the good modules are discarded leaving one remaining good module. The reliability of the TMR/Simplex system if an ideal voter is assumed is given by [22]:

$$R_{TMR/Simplex} = 1.5R_m - 0.5R_m^3 \tag{2}$$

Figure 1 contrasts the reliability of Simplex, TMR, and TMR/Simplex systems versus the normalized mission time (time/MTTFsimplex). As can be seen from Figure 1, TMR is better than Simplex until 0.7 MTTFsimplex; however, TMR/Simplex is always better than either TMR or Simplex alone.

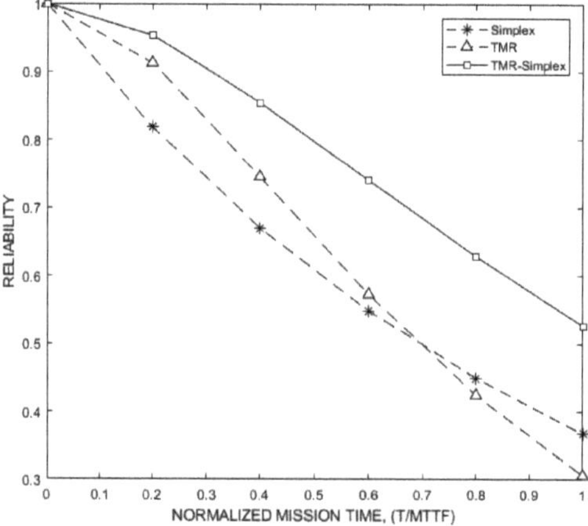

Figure 1. Reliability versus normalized mission time for Simplex, TMR, and TMR/Simplex systems.

As a measure of radiation hardness indicator, a parameter that has been found valuable for evaluating reliable systems is the Reliability Improvement Factor (RIF) [22,23]. This is defined as the ratio of the probability of failure of the non-redundant system to that of the redundant system. If R_N and R_R are the reliabilities of the non-redundant and the redundant systems, respectively, for a given mission time, and at a given radiation level, then:

$$RIF = \frac{1 - R_N}{1 - R_R} \tag{3}$$

Therefore, assuming Simplex module failure rate, $\lambda = 5\%$, the RIF over the Simplex system of the TMR method is 7.25, whereas the RIF of the TMR/Simplex scheme over the Simplex system is 14.3.

The literature reports the output of research showing alternative methods to TMR or TMR/Simplex, which have better reliability and reduced overhead costs [24,25]. However, excluding Lima's [26] hybrid method, which has been shown to provide reliability benefits equivalent to those of TMR at a lower cost, no other method has been reported in the literature.

2. The Proposed Technique

The TMR/Simplex scheme provides higher reliability than either the TMR or Simplex system alone, consequently increasing the mission time for which the scheme can be used. However, it abandons a working resource and completely fails as soon as the selected good module fails; thus, another technique is desired that will not discard a working resource or cause a complete system failure as soon as the second module fails. In addition, the scheme would be suitable for a radiation-tolerant digital controller design.

In an effort to address the above research questions, a new redundancy architecture is proposed, as shown in Figure 2. In this architecture, a hybrid redundancy scheme that combines Spatial and Temporal redundancies is used to design a high-reliability redundancy scheme that mitigates the problems encountered with the ordinary TMR/Simplex technique.

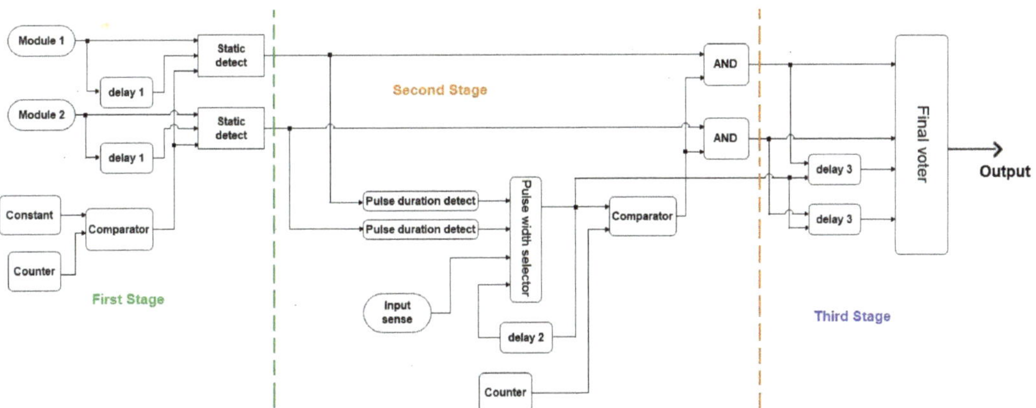

Figure 2. Implementation of the proposed technique for the two-module case.

The motivation for the proposed technique rests on the fact that radiation-induced faults result in one or more of the following effects on the DC-DC converter PWM controller output [27–29]:

1. A change in pulse duration(s) of the PWM controller output (for one or more cycles).
2. A loss of pulse (due to complete failure of the PWM controller; mostly assumed to be due to the permanent change in output to logic-low or 0).
3. Missing pulses (which occur because the PWM controller's output is stuck at logic-high or logic-low for one or more cycles).

For a given input voltage, the operating duty cycle and, consequently, the logic-high duration of the pulse generated by the PWM controller can be known. The knowledge of this pulse duration can be used to detect the occurrences of the three radiation-induced fault categories mentioned above and mask them so that their effect does not alter the correct output state.

To illustrate the concept, a two-module implementation of the proposed technique is shown in Figure 2. However, the numbers of possible paralleled redundant elements are limited only by other constraints, such as space and power requirements of a given design; otherwise, 2, 3, 4, 5, 6, 7, etc., redundant elements, irrespective of being an odd or even number of elements, can be paralleled to reach the required level of reliability. Consequently, for radiation hardening half-duty limited DC-DC converters, the reliability obtained from the proposed technique outweighs most, if not all, of the ordinary modular redundancy techniques.

Figure 2 shows three stages of the proposed voter. The functioning of each stage is as follows:

2.1. First Stage

This stage consists of a counter, a comparator, a constant block, two delay blocks, and two first-stage voters. In this stage, the following faults are detected,

1. Faults that result in a PWM pulse duration change larger than the maximum duty-limit.
2. Faults that result in being stuck at logic-high or stuck at logic-low for one or more PWM cycles.

If the above two fault categories are detected, these controllers' outputs are replaced with a low-duration pulse of the same frequency. The actual masking of these fault types happens at the third stage. The first-stage voter (static detect in Figure 2) inputs the following signals:

1. Each module's PWM pulse output;
2. Each module's PWM pulse output delayed by the maximum duty-limit used;
3. A low-duration PWM pulse of the same frequency.

In this stage, if each module's PWM output pulse is free from the faults categorized above or if radiation causes a pulse-duration change smaller than the fixed maximum duty-limit used, then a comparison of that module's PWM output pulse and its fixed maximum duty-limit delayed counterpart should result in a difference, as shown in Figure 3a (upper-blue and middle-red). If so, the first-stage voter propagates that pulse to the next stage as shown in Figure 3a (lower-brown). However, if radiation causes a change in the pulse-duration larger than the maximum duty-limit used, then the first-stage voter passes that pulse only for the duration of time for which that pulse and its maximum duty delayed counterpart have dissimilarity; otherwise, the low-duration pulse is passed as shown in Figure 3b. Furthermore, if radiation causes a fault of being permanently or temporarily stuck at logic-low or logic-high, a comparison with its fixed maximum duty-limit delayed equivalent will not result in a difference. In this case, the first-stage voter will replace that particular module's output with a low-duration pulse with the same frequency, as shown in Figure 3c,d, respectively. The first-stage voter pseudocode is shown in Figure 4.

Note that this stage detects either permanent or temporary stuck at logic-low or stuck at logic-high faults that persist for one or more PWM cycles, and faults that result in a change in PWM pulse duration that is larger than the maximum duty-limit used. However, radiation-induced faults that result in a change in PWM pulse duration that is smaller than the maximum duty-limit are not detected in this stage.

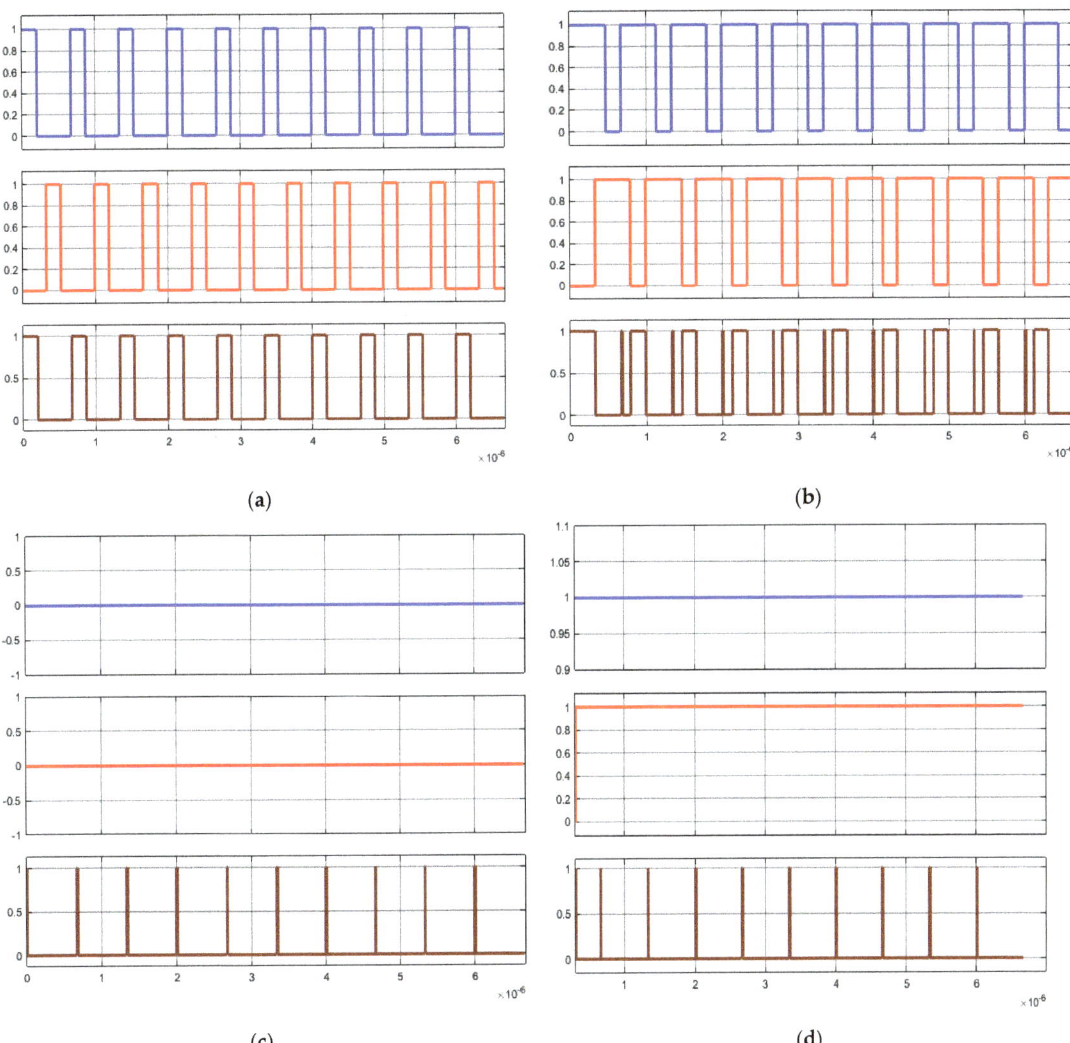

Figure 3. First-stage detection process: (**a**) A 1.5 MHz 30% actual duty PWM pulse (upper-blue), its 48% pulse-duration delayed counterpart (middle-red) and resultant first-stage voter output PWM pulse (lower-brown); (**b**) a faulty 1.5 MHz 70% duty (larger than the maximum duty, max-duty = 48%), PWM pulse (upper-blue), its 48% pulse-duration delayed counterpart (middle-red) and resultant first-stage voter output PWM pulse (lower-brown); (**c**) a faulty 1.5 MHz 0% duty (stuck at logic-low fault), PWM pulse (upper-blue), its 48% pulse-duration delayed counterpart (middle-red) and resultant first-stage voter output PWM pulse (lower-brown); (**d**) a faulty 1.5 MHz 100% duty (stuck at logic-high fault), PWM pulse (upper-blue), its 48% pulse-duration delayed counterpart (middle-red) and resultant first-stage voter output PWM pulse (lower-brown).

INPUT (PWM_PULSE, PWM_PULSE_DELAYED, LOW_DURATION_PWM_PULSE)

{ IF (PWM_PULSE is_diffferent_from **PWM_PULSE_DELAYED)**

    OUTPUT is_equal_to **PWM_PULSE;** // no fault or fault with smaller duration than max duty-limit.

ELSE

    OUTPUT is_equal_to **LOW_DURATION_PWM_PULSE;** // fault detected replace with low duration pulse.

}

Figure 4. First-stage static detect block pseudocode.

2.2. Second Stage

This stage consists of a counter, a comparator, a delay block, two pulse-duration detection algorithm blocks, and two two-input AND blocks. This stage performs the following functions:

1. Detects the pulse-durations of each inputted pulse and decides the current actual pulse duration.
2. Detects and rejects faults that result in a PWM pulse duration change smaller than the maximum duty-limit but larger than the actual PWM pulse duration

In this stage, the pulse durations of each inputted pulse are detected, and the actual pulse duration is selected and used to generate a pulse that will be ANDED with the first stage's outputs. The ANDING process will allow, passing to the third stage, only those PWM pulses that have equal or smaller pulse-durations than that of the selected actual pulse duration.

Note that the previous correct PWM cycle's duty-value, and the current input voltage value (that is, the fact that the product of the input voltage and primary turn-on time is almost a constant value, no matter how fast the input voltage changes), is used to select the correct pulse duration in the second-stage voter (pulse width selector block in Figure 2). Therefore, the only faults that can pass through this stage are those that result in smaller pulse durations than that of the actual pulse duration.

The second stage works by inputting:

1. The outputs from the pulse-duration detectors;
2. The previous PWM cycle's duty-value;
3. The current input voltage–output voltage relation, that is, current duty-value calculated using the equation:

$$duty = \frac{N \times \text{Output voltage}}{\text{Current input voltage}} \times NPWM \tag{4}$$

where N is the turn-ratio and NPWM = 2^8 = 256 for the 8-bit DPWM used in the article. Here, because N, NPWM, and output voltage are constants, only the current input voltage value is sensed by the input sensing circuit.

The pseudocode for the second-stage pulse width selector block in Figure 2 is shown in Figure 5:

```
INPUT (PULSE_DURATION_1, PULSE_DURATION_2, PREVIOUS_DUTY, DUTY)

{ IF (PULSE_DURATION_1 is_equal_to PULSE_DURATION_2)

    OUTPUT is_equal_to PULSE_DURATION_1; // no fault.

ELSE_IF ((PREVIOUS_DUTY is_less_or_equal_to PULSE_DURATION_1 + TOLLERANCE)

 && (PREVIOUS_DUTY is_greater_or_equal_to PULSE_DURATION_1 – TOLLERANCE))

    OUTPUT is_equal_to PULSE_DURATION_1; // module 1 is fault free, take it as correct actual duty.

ELSE_IF ((PREVIOUS_DUTY is_less_or_equal_to PULSE_DURATION_2 + TOLLERANCE)

 && (PREVIOUS_DUTY is_greater_or_equal_to PULSE_DURATION_2 – TOLLERANCE))

    OUTPUT is_equal_to PULSE_DURATION_2; // module 2 is fault free, take it as correct actual duty.

ELSE

    OUTPUT is_equal_to DUTY; // only for the moment when large input/output change brings large actual

    duty change from the previous cycle's value.

}
```

Figure 5. Second-stage pulse width selector block pseudocode.

The TOLERANCE value is based on the allowable output voltage variation/tolerance. The value used in the article is 2 clock durations, which corresponds to an allowable duty-value variation of 0.78% or output voltage variation of a maximum of 140 mV above or below the nominal 4 V value (maximum of 140 mV variation occurs at the largest input voltage). The TOLERANCE value can be tightened if required.

Figure 6 below shows the simulation runs of the three-module implementation of the proposed technique. In the figures, initially the three-module system is running with the actual duty-value of 30%. Then, after approximately 0.46 milliseconds, the first module is switched to a duty-value of 10% to emulate a radiation-induced fault (Figure 6a); after approximately 0.8925 milliseconds, the second module is switched to a duty-value of 10% (Figure 6b). Figure 6c shows outputs after AND blocks with the first module switched to a duty-value of 80% after approximately 0.4975 milliseconds of the simulation run, and Figure 6d shows the outputs with the second module switched to a duty-value of 80% after approximately 0.8275 milliseconds of the simulation run. In Figure 6c,d, because faulty pulses that have larger pulse durations than the maximum duty-value are masked by the first stage, the outputs after AND blocks are constant zero-duration pulses which can easily be masked by the third-stage voter. In all figures, the bottom pulse graph shows the resultant actual PWM pulse output after the third stage during the simulation runs. This shows that the failure(s) of one or two module(s) is masked by the two or single fault-free remaining module(s).

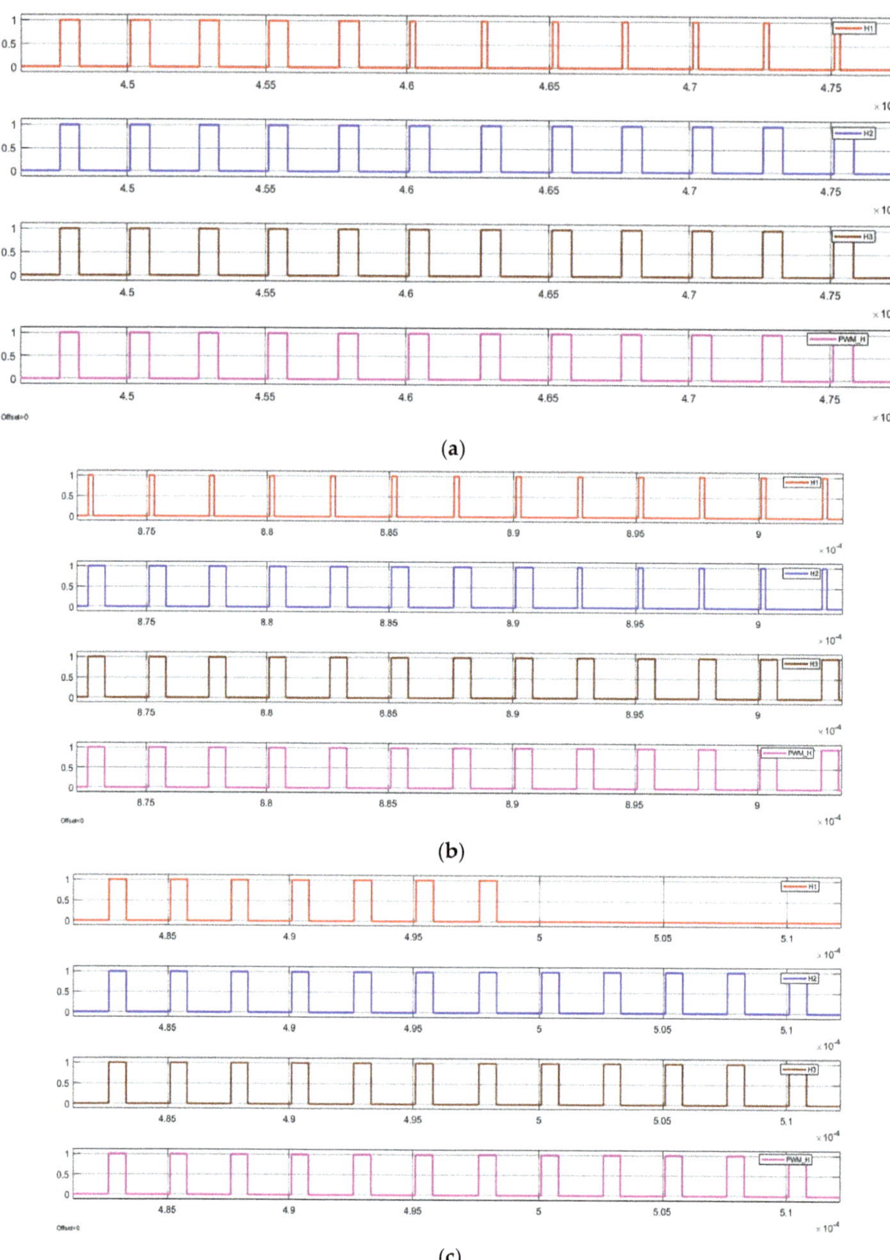

(**a**)

(**b**)

(**c**)

Figure 6. *Cont.*

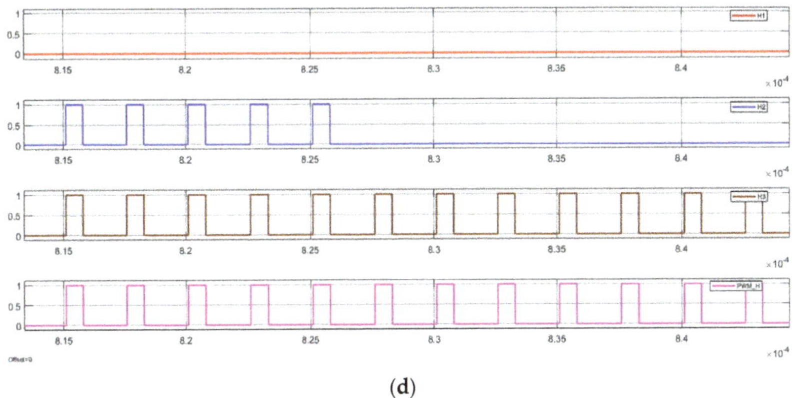

(d)

Figure 6. (**a**) Outputs after AND blocks and the final voter when module 1 is forced to output 10% duty to emulate a sudden change in pulse duration. (**b**) Outputs after AND blocks and the final voter when module 2 is also forced to output 10% duty to emulate a sudden change in pulse duration. (**c**) Outputs after AND blocks and the final voter when module 1 is forced to output 80% duty to emulate a sudden change in pulse duration. (**d**) Outputs after AND blocks and the final voter when module 2 is also forced to output 80% duty to emulate a sudden change in pulse duration.

2.3. Third Stage

This stage consists of two run-time delay blocks and the final voter. In this stage, a run-time, dynamically generated, delay is used to detect and reject smaller pulse-duration faulty pulses that have passed through the second stage.

The pseudocode for the third stage is similar to that of the first stage, except that in the third stage, no low-duration pulse is required as a replacement for the faulty pulses; in addition, the delay duration is dynamically determined at the second stage and, thus, is not constant. The final voter found in this stage determines the final correct actual PWM pulse by inputting the two redundant modules' PWM outputs and their actual duty-value delayed counterparts.

The pseudocode for the third stage is shown in Figure 7:

```
INPUT (PULSE_OF_MODULE_1, PULSE_OF_MODULE_2, ACTUAL_PWM_DELAYED_PULSE_1,

ACTUAL_PWM_DELAYED_PULSE_2)

{ IF (PULSE_OF_MODULE_1 is_different_from ACTUAL_PWM_DELAYED_PULSE_1)

        OUTPUT is_equal_to PULSE_OF_MODULE_1; // module 1 is fault free, take it as the final correct actual duty.

ELSE_IF ( PULSE_OF_MODULE_2 is_different_from ACTUAL_PWM_DELAYED_PULSE_2)

        OUTPUT is_equal_to PULSE_OF_MODULE_2; // module 2 is fault free, take it as the final correct actual duty.

ELSE

        OUTPUT is_equal_to 0; // total failure.

}
```

Figure 7. Third-stage final voter block pseudocode.

As can be observed from Figure 8, although the actual pulse-duration delayed pulse (Figure 8d) results in a dissimilarity compared to the actual PWM pulse (Figure 8b), at each clock cycle in the PWM cycle the smaller pulse-duration faulty pulse (Figure 8a) is not different from its actual pulse-duration delayed equivalent (Figure 8c), which can easily be detected and rejected by the final voter.

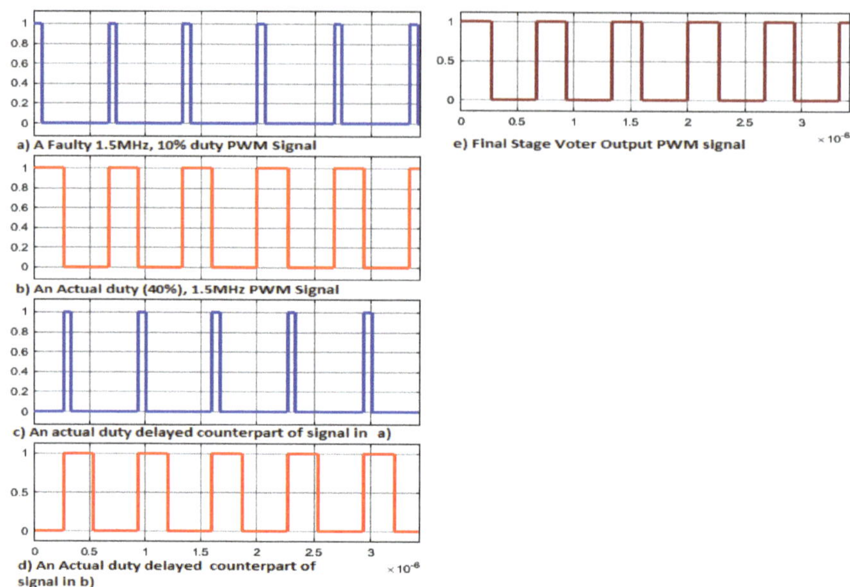

Figure 8. Third-stage detection and rejection process.

3. Reliability Analysis of the Two-Modules Implementation Case

To conduct an effective evaluation of the dependability of a given fault tolerant system, a measure of reliability is required. Although the true reliability obtained from the proposed technique is dependent on the number of paralleled redundant modules, and increases with the number of modules, the reliability of the two-module implementation case, shown in Figure 2, can be calculated. Assuming Rm to be the reliability of one of the modules (Simplex system), the reliability expression of the proposed technique, if an ideal voter is assumed, can be derived from the following equations:

$$R_{\text{two modules}} = \text{Probability of both modules are functioning} + \text{probability of only one of the modules is functioning} \tag{5}$$

$$R_{\text{two modules}} = B(2:2) + B(2:1) = \binom{2}{2} R_m^2 (1 - R_m)^0 + \binom{2}{1} R_m^1 (1 - R_m)^1 \tag{6}$$

$$R_{\text{two modules}} = 2R_m - R_m^2 = 2e^{-\lambda t} - e^{-2\lambda t} \tag{7}$$

Therefore, the RIF of the proposed two-module implementation over the Simplex system is 20.8. This represents a 2.87-fold and 1.46-fold improvement in RIF over the ordinary TMR and TMR/Simplex schemes, respectively, for the same system.

The graph in Figure 9 compares the reliability of Simplex, TMR, TMR/Simplex, and the proposed technique implementations with two and three modules versus the normalized mission time (time/MTTFsimplex).

As can be seen from Figure 9, the proposed technique provides the best reliability for all $t \geq 0$, compared to either TMR or TMR/Simplex methods, which makes it suitable for applications to relatively longer mission times. The graph also accentuates the claim that the reliability obtained from the proposed technique increases with the increase in the number of paralleled redundant elements.

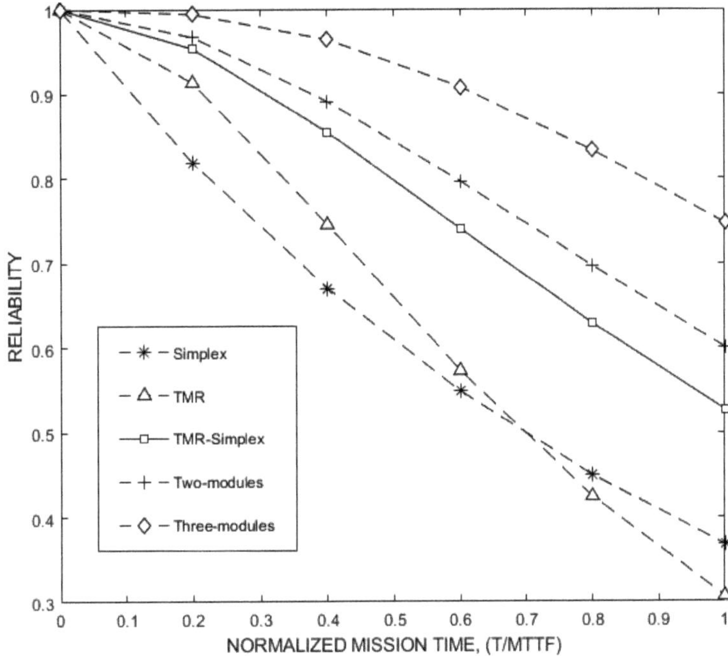

Figure 9. Reliability versus normalized mission time for Simplex, TMR, TMR/Simplex, and the proposed method for two- and three-module cases.

4. Case Study

The Dual-Switch Forward Converter

Among the different topologies of DC-DC converters, those considered suitable for applications in radiation environments are required to provide insulation between input and output of the converter, and the power switch arrangement should not cause a short circuit at the input of the converter in the case of a fault caused by radiation. Therefore, Forward or Flyback converters are frequently selected [30].

The Dual-Switch Forward converter, like the typical Single-Switch Forward converter, is derived from the Buck converter topology. The key difference between a Forward converter and a Buck converter is that a transformer is introduced in the Forward converter. The transformer creates the input–output separation, and the turn ratio offers a means to adjust the duty cycle for the particular input and output voltage requirements of the application. Figure 10 illustrates the Dual-Switch Forward converter topology. The circuit consists of an input capacitor CIN, two switches QH and QL, clamp diodes DH and DL, a power transformer T1, rectifier diodes D1 and D2, an inductor Lo, and a capacitor Co. With a somewhat higher cost, the rectifier diodes D1 and D2 on the secondary side of the power transformer can be replaced with synchronous rectifier switches to improve efficiency in applications with relatively low output voltage.

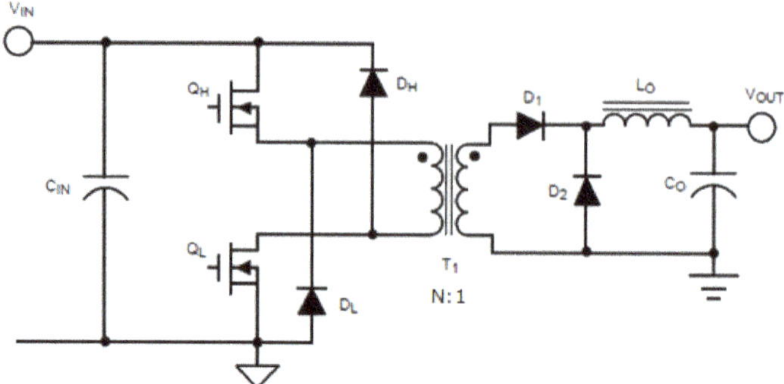

Figure 10. Dual-Switch Forward converter topology.

Figure 11 shows the two operational modes of the Dual-Switch Forward converter. During operation, the two switches are turned ON and OFF simultaneously. The output voltage is regulated by modulating the duty cycle of the switches. The relationship between the input voltage VIN, output voltage VOUT, duty cycle D, rectifier diode forward-drop VF, and transformer turns ratio N is defined by the following equation:

$$D = \frac{N \times (V_{OUT} + V_F)}{V_{IN}} \tag{8}$$

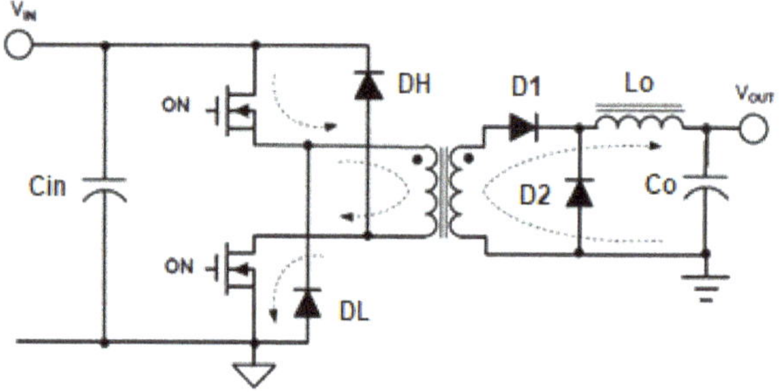

(**a**) Current paths when switches are turned on.

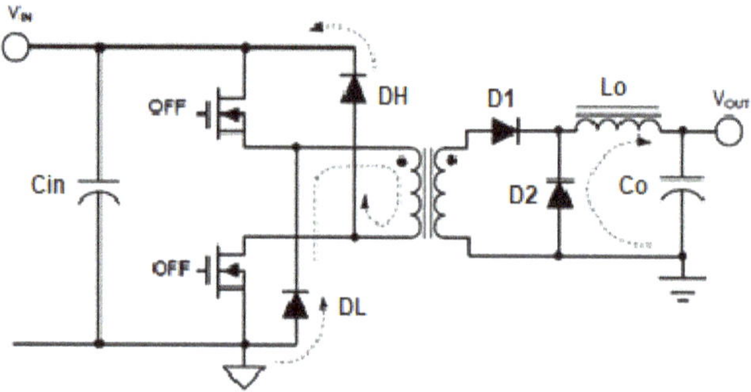

(**b**) Current paths when switches are turned off.

Figure 11. Operating modes of Dual-Switch Forward converter.

When the two switches are turned ON, as shown in Figure 11a, the input voltage is applied to the primary power transformer. Consequently, the transformer core is magnetized, and the power flows to the secondary side circuit through the transformer coupling. When the two switches are OFF, as shown in Figure 11b, the flow of power to the primary is cut off. The voltage across the primary winding is reversed due to the residual magnetizing inductance of the transformer, forcing the two clamp diodes DH and DL to conduct. This effectively clamps the switches' voltage to the input voltage, and applies the input voltage in reversed polarity to the power transformer primary winding to demagnetize and reset the transformer.

The primary of the transformer receives the voltage of the nearly equal magnitude but opposite polarities during the ON and OFF period of the power switches. The maximum duty cycle should be limited to less than 50% to ensure the volt-second balance between the magnetizing and demagnetizing intervals, so that the Dual-Switch Forward converter always achieves a complete reset of the power transformer during each switching cycle.

5. Design Parameters of the Converter

Design specifications of the Dual-Switch Forward converter are shown in the Table 1 below.

Table 1. Design parameters of the Dual-Switch Forward converter.

Parameter	Rating Value
DC Input Voltage (Vin) range	80–144 V
Turn ratio, N	8
Output Voltage (Vout)	4 V
Output Current (Io) range	2–20 A
Inductor (L), ESL	1 µH, 8 mΩ
Capacitor (C), ESR	13 µF, 15 mΩ
Load (R) range	0.2–2 Ω
Switching Frequency (Fsw)	1.5 MHz
Output Power (Po)	80 Watts
Maximum duty cycle (Dmax)	0.48
Efficiency (η)	>90%

A voltage-mode PWM controller (VMC) for the Dual-Switch Forward converter was designed in the analog domain using the MATLAB control system toolbox. The designed PID compensator has a gain margin of 11.2 dB and a phase margin of 54.2 degrees.

$$G_{Comp}(s) = \frac{7.713\,e^{-8}\left(s + 4.33e^5\right)^2}{s} \tag{9}$$

The designed analog compensator was then converted to its equivalent digital form using the bilinear transformation. The final digital PID compensator transfer function is given by:

$$G_c(z) = \frac{2.41e^{-2} - 3.74e^{-2}z^{-1} + 1.45e^{-2}z^{-2}}{1 - z^{-1}} \tag{10}$$

6. Reliability, Hardware Resources, and Mean-Time-To-Failure (MTTF) Comparisons

The Mean-Time-to-Failure (MTTF) [21] and reliability comparisons of the methods are shown in Table 2. As can be seen from Table 2, the proposed technique is superior for relatively longer mission time applications than either TMR or TMR/Simplex techniques.

Table 2. Reliability and MTTF comparison.

Methods	Reliability and MTTF (λ=5%)		
	MTTF (in years)	Reliability for Mission Time t	
		t = 1 year	RIF
Simplex	20	0.95	1
TMR	145	0.9931	7.25
TMR/Simplex	286	0.9965	14.30
Proposed Technique (Two Modules)	417	0.9976	20.83
Proposed Technique (Three Modules)	10,000	0.9999	500

Reliability is defined as the probability of not failing in a particular environment for a particular mission time. The MTTF is derivable from the reliability of a given system. The details are given in reference [21].

Respective MTTF values indicate an average lifespan before the first failure of the system obtainable using each respective technique.

The Reliability Improvement Factor (RIF), also called the Reliability Improvement Index (RII) in some literature, is a measure of the relative advantage of one redundancy technique over the other with respect to the unmitigated system.

Table 3 shows analysis results, for each technique, of hardware resources after synthesis.

Table 3. Hardware resource comparisons.

-	DSP (80)	LUT (17,600)	Registers (35,200)
TMR\TMR/Simplex	3	1500	1467
Proposed Technique (Two Modules)	3	1317	1536
Proposed Technique (Three Modules)	4	1793	2094

7. Testing

To verify the functionality of the proposed technique, the digital controller for the Dual-Switch Forward converter was implemented in the zynq-7000 (zybo) board using a Xilinx system generator and MATLAB/Simulink. The hardware co-simulation block was generated.

A Hardware co-simulation setup was used to emulate radiation-induced faults during simulation. In this setup, the pulse generator block of MATLAB/Simulink was used to emulate (or inject) the pulse-duration changes caused by radiation-induced faults during simulation, and the loss of pulses or stuck at logic-high or stuck at logic-low faults were emulated using a constant block.

Two experiments were designed. Experiment 1 was used to test the injections of the three radiation-induced fault categories in the presence of input disturbances, whereas experiment 2 was used to test the injections of the three fault categories in the presence of output disturbances. Table 4 shows the duty-value changes injected to emulate the radiation-induced faults, and Figure 12 shows the synthesizable fault models inserted into the desired places in the VHDL design to emulate the radiation-induced faults.

Table 4. Duty-value change used for different experiment sequences.

Fault Type	Stuck at Logic-Low or Permanent Failure	Duty-Value Changes					Stuck at Logic-High
Sequence No	1	2	3	4	5	6	7
Duty-Value (%)	0	10	40	60	80	90	100

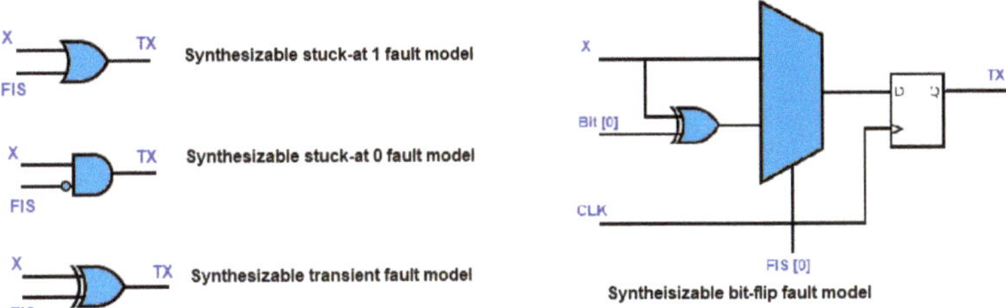

Figure 12. Synthesizable fault models.

7.1. Experiment 1

In experiment 1, the simulation interval was initially set at 2 milliseconds, the input DC-bus voltage was made to switch between 144 V and 128 V at a 0.3 millisecond interval, and the output load was fixed at 0.2 Ω. Then the following sequence of events was executed:

1. Sequence 1: Stuck at logic-low for Multiple PWM cycles

    a. At t = 0 ms, the simulation was started with the two controller modules outputting the same actual PWM pulse, that is, actual-duty = $8 \times 4/144 = 0.22$, (a small duty-value increase due to losses was ignored and the ideal diode was used, so $V_F = 0$, as assumed in Equation (7)).

    b. At t = 0.2 ms, the first controller module output was switched to duty = 0.

    c. Starting from t = 0.3 ms, the periodic input disturbance was injected and repeated at 0.3 ms intervals of switching between 144 V and 128 V until the simulation was complete (note that the actual duty-value changes from 0.22 to 0.25 when the input DC-bus voltage changes from 144 V to 128 V).

    d. At t = 0.5 ms, the first module was restored (switched back to the fist controller's output).

    e. At t = 0.7 ms, the second controller module output was switched to duty = 0.

    f. At t = 1 ms, the second module was restored (switched back to the second controller's output).

    g. At t = 1.3 ms, the first controller module output was switched to duty = 0.

    h. At t = 1.6 ms, the first module was restored (switched back to the fist controller's output).

The simulation was repeated for all duty-value changes shown in Table 4 to emulate other fault types.

7.2. Experiment 2

In experiment 2, the same procedure and simulation interval was used as in experiment 1, but the input DC-bus voltage was fixed at 144 V and the output load fixed part was set to 0.8 Ω; in addition, the cyclic load current demand switched between 0 A and 2.5 A so that the total load current demand switched between 5 A and 7.5 A at 0.3 millisecond intervals starting from 0.3 milliseconds after the simulation started.

7.3. Testing Using Synthesizable Fault Models

In this case, synthesizable fault models from [31] were used to inject different fault types into the required places in the VHDL design. The different types of synthesizable faults are shown in Figure 12.

When injecting the above faults at the required locations in the design, the Fault Injection System (FIS) wire, in Figure 12, plays a major role. The respective faults would be activated if FIS takes a value of 1, and the faults would become inactive if it takes the value 0. Figure 13 shows the schematic of the system after synthesis with synthesizable bit-flip fault models inserted at the outputs of each controller for a three-module implementation of the technique.

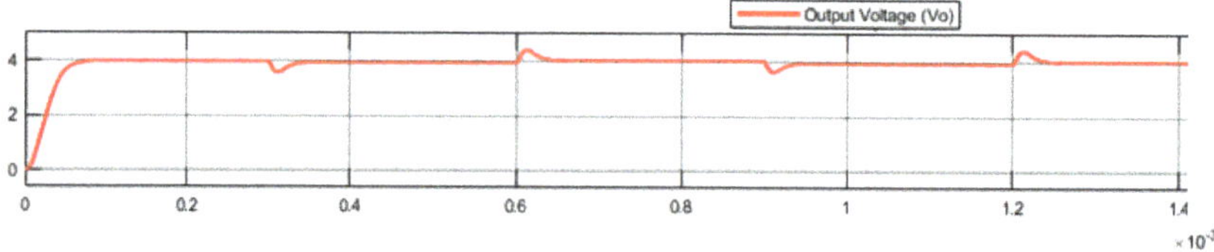

Figure 13. Schematic after synthesis showing the location of the insertion of bit-flip synthesizable fault models.

When testing with the synthesizable fault models, similar responses of the converter to those in the cases of experiments 1 and experiment 2 (shown in Figures 14 and 15) were observed in the presence of input disturbance (experiment 1) and output disturbance (experiment 2). This proves the effectiveness of the proposed method.

Figure 14. Converter output voltage response during input disturbances in the presence of radiation fault injection (experiment 1).

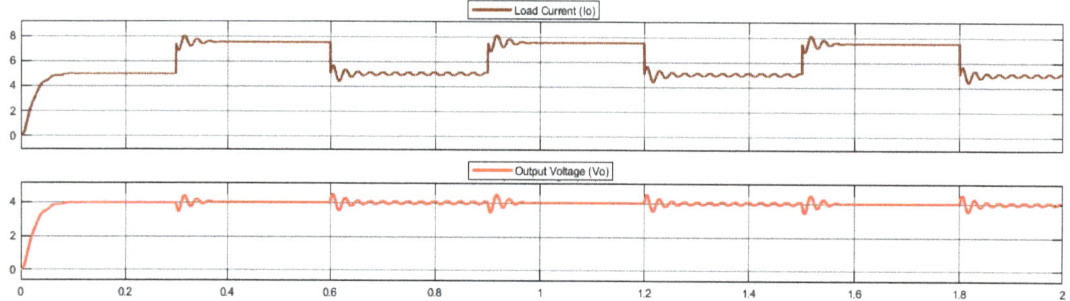

Figure 15. Converter output voltage and current responses during load disturbances in the presence of radiation fault injection (experiment 2).

8. Experimental Results

The converter output responses are shown in Figure 14 for experiment 1, and in Figure 15 for experiment 2. As can be seen from Figures 14 and 15, the converter tolerates the three radiation-induced fault categories previously discussed in the presence of input or output disturbances. Similar converter responses were observed for the different duty-value changes shown in Table 3, and for the injections of the three radiation-induced fault categories in the presence of input or output disturbances.

Figure 16 shows the hardware co-simulation setup used during the experiments.

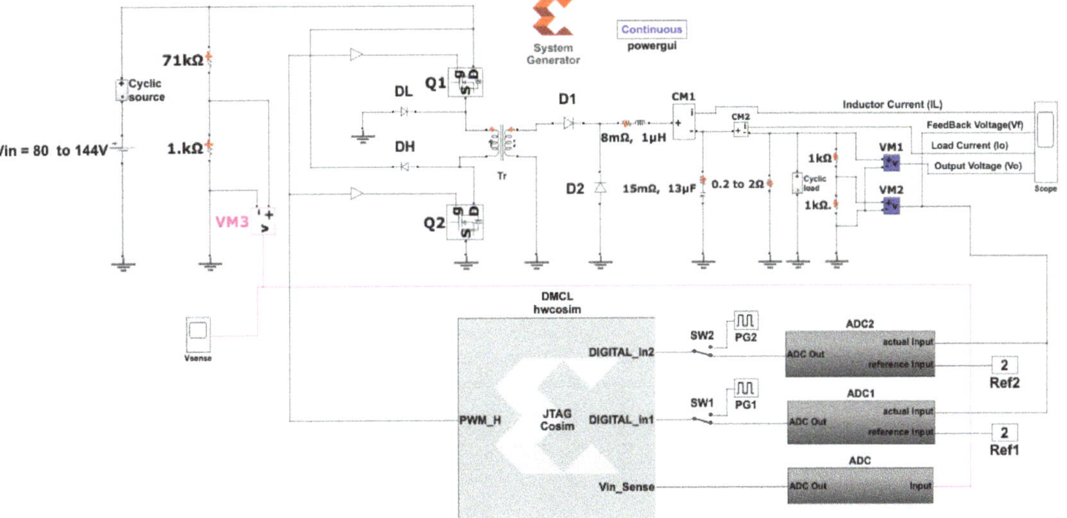

Figure 16. Hardware co-simulation setup.

9. Discussion

The proposed method is applicable for radiation hardening of half-duty limited DC-DC converters and inverters or similar circuits and/or applications; examples include isolated Dual-Switch Forward and Flyback DC-DC converters. The main limitation is that the method initially adds a delay of two PWM cycles (approximately 1.33 µs in this paper) in the control loop due to the pulse-duration detection algorithms.

The main advantage of the technique is that it can be used to parallel any number of redundant modules, irrespective of being even or odd numbers, with a significant increase in reliability with the number of paralleled redundant modules. Furthermore, the technique continues to function even if only one module is free from radiation-induced faults. The two-module implementation of the technique consumes slightly less resources (LUT) compared to the TMR implementation. Each redundant module's ADC implementation requires one DSP; thus, TMR uses three DSPs, whereas the proposed method's two-module and three-module implementations use three and four DSPs, respectively. One more DSP is required for the implementation of the ADC in the input voltage sensing circuit. Overall, the method provides significantly higher reliability, in addition to efficiently using resources. The three-module implementation consumes less than 2% more resources (LUT, registers), and uses one more DSP compared to the TMR, but the increase in reliability is significant.

10. Conclusions

In this paper, a high-reliability hybrid redundancy technique as an alternative to the regular TMR or TMR/Simplex schemes for radiation hardening of half-duty limited DC-DC converters is presented. The technique provides the highest reliability compared to TMR and TMR/Simplex schemes. For the two-module implementation case presented in this paper, the method provides 2.87-fold and 1.46-fold RIF over the Simplex system, compared to TMR and TMR/Simplex techniques, respectively.

The technique can be used to parallel any number of redundant modules, irrespective of being even or odd numbers, with a significant increase in reliability with the number of paralleled redundant modules. The technique can be used for longer mission time applications than both TMR and TMR/Simplex techniques. The technique can be used for half-duty limited DC-DC converters or similar circuits in space systems, and/or in nuclear or high energy physics facilities.

Author Contributions: Conceptualization, S.M.B.; methodology, S.M.B.; software, S.M.B.; validation, S.M.B. and P.L.; formal analysis, S.M.B.; investigation, S.M.B.; resources, S.M.B. and P.L.; writing—original draft preparation, S.M.B.; writing—review and editing, S.M.B. and P.L.; visualization, S.M.B.; supervision, P.L., G.B., V.D.S. and B.K.; project administration, P.L. and S.M.B. All authors have read and agreed to the published version of the manuscript.

Funding: This research was supported by the Home-Grown PhD Program (HGPP) funded by the Ethiopian Ministry of Education.

Conflicts of Interest: The authors declare no conflict of interest.

References

1. Pratt, B.; Caffrey, M.; Graham, P.; Morgan, K.; Wirthlin, M. Improving FPGA design robustness with partial TMR. In Proceedings of the 2006 IEEE International Reliability Physics Symposium Proceedings, San Jose, CA, USA, 26–30 March 2006; pp. 226–232.
2. Bernardi, P.; Reorda, M.S.; Sterpone, L.; Violante, M. On the evaluation of SEU sensitiveness in SRAM-based FPGAs. In Proceedings of the 10th IEEE International On-Line Testing Symposium, Funchal, Portugal, 14 July 2004; pp. 115–120.
3. Fay, D.; Shye, A.; Bhattacharya, S.; Connors, D.A.; Wichmann, S. An adaptive fault-tolerant memory system for FPGA-based architectures in the space environment. In Proceedings of the Second NASA/ESA Conference on Adaptive Hardware and Systems (AHS 2007), Scotland, UK, 5–8 August 2007; pp. 250–257.
4. Jacobs, A.; George, A.D.; Cieslewski, G. Reconfigurable fault tolerance: A framework for environmentally adaptive fault mitigation in space. In Proceedings of the 2009 International Conference on Field Programmable Logic and Applications, Prague, Czech Republic, 31 August 2009; pp. 199–204.
5. Aguiar, Y.Q.; Wrobel, F.; Autran, J.-L.; Leroux, P.; Saigné, F.; Pouget, V.; Touboul, A.D. Design exploration of majority voter architectures based on the signal probability for TMR strategy optimization in space applications. *Microelectron. Reliab.* **2020**, *114*, 113877. [CrossRef]
6. Janson, K.; Treudler, C.J.; Hollstein, T.; Raik, J.; Jenihhin, M.; Fey, G. Software-level tmr approach for on-board data processing in space applications. In Proceedings of the 2018 IEEE 21st International Symposium on Design and Diagnostics of Electronic Circuits & Systems (DDECS), Budapest, Hungary, 25–27 April 2018; pp. 147–152.
7. Padmanabhan, B.; Mastorakis, N.E. Power, delay and area comparisons of majority voters relevant to TMR architectures. *arXiv* **2016**, arXiv:1603.07964.
8. Agiakatsikas, D.; Nguyen, N.T.H.; Zhao, Z.; Wu, T.; Cetin, E.; Diessel, O.; Gong, L. Reconfiguration control networks for TMR systems with module-based recovery. In Proceedings of the 2016 IEEE 24th Annual International Symposium on Field-Programmable Custom Computing Machines (FCCM), Washington, DC, USA, 1–3 May 2016; pp. 88–91.
9. Siddiqui, K.S.; Baig, M.A. FRAM based TMR (triple modular redundancy) for fault tolerance implementation. In Proceedings of the 2011 International Conference on Information and Communication Technologies, Azerbaijan, Baku, 12–14 October 2011; pp. 1–5.
10. Hudson, S.; Sundar, R.S.S.; Koppu, S. Fault control using triple modular redundancy (TMR). In *Progress in Computing, Analytics and Networking*; Springer: Singapore, 2018; pp. 471–480.
11. Lee, G.; Agiakatsikas, D.; Wu, T.; Cetin, E.; Diessel, O. TLegUp: A TMR code generation tool for SRAM-based FPGA applications using HLS. In Proceedings of the 2017 IEEE 25th Annual International Symposium on Field-Programmable Custom Computing Machines (FCCM), Napa, CA, USA, 30 April–2 May 2017; pp. 129–132.
12. Bernardeschi, C.; Cassano, L.; Domenici, A. SRAM-based FPGA systems for safety-critical applications: A survey on design standards and proposed methodologies. *J. Comput. Sci. Technol.* **2015**, *30*, 373–390. [CrossRef]
13. Li, T.; Liu, H.; Yang, H. Design and characterization of SEU hardened circuits for SRAM-based FPGA. *IEEE Trans. Very Large-Scale Integr. (VLSI) Syst.* **2019**, *27*, 1276–1283. [CrossRef]
14. Tarrillo, J.; Tonfat, J.; Tambara, L.; Kastensmidt, F.L.; Reis, R. Multiple fault injection platform for SRAM-based FPGA based on ground-level radiation experiments. In Proceedings of the 2015 16th Latin-American Test Symposium (LATS), Puerto Vallarta, Mexico, 25–27 March 2015; pp. 1–6.
15. Schmidt, F.H., Jr. Fault Tolerant Design Implementation on Radiation Hardened by Design SRAM-Based FPGAs. Ph.D. Thesis, Massachusetts Institute of Technology, Cambridge, MA, USA, 2013.

16. Quinn, H.; Wirthlin, M. Validation techniques for fault emulation of SRAM-based FPGAs. *IEEE Trans. Nuclear Sci.* **2015**, *62*, 1487–1500. [CrossRef]

17. Mousavi, M.; Pourshaghaghi, H.R.; Tahghighi, M.; Jordans, R.; Corporaal, H. A generic methodology to compute design sensitivity to SEU in SRAM-based FPGA. In Proceedings of the 2018 21st Euromicro Conference on Digital System Design (DSD), Prague, Czech Republic, 29–31 August 2018; pp. 221–228.

18. Giordano, R.; Perrella, S.; Izzo, V.; Milluzzo, G.; Aloisio, A. Redundant-configuration scrubbing of SRAM-based FPGAs. *IEEE Trans. Nuclear Sci.* **2017**, *64*, 2497–2504. [CrossRef]

19. Jung, S.; Choi, J.P. Predicting system failure rates of SRAM-based FPGA on-board processors in space radiation environments. *Reliab. Eng. Syst. Saf.* **2019**, *183*, 374–386. [CrossRef]

20. Steckert, J.; Skoczen, A. Design of FPGA-based radiation tolerant quench detectors for LHC. *J. Instrum.* **2017**, *12*, T04005. [CrossRef]

21. Banteywalu, S.; Khan, B.; De Smedt, V.; Leroux, P. A Novel Modular Radiation Hardening Approach Applied to a Synchronous Buck Converter. *Electronics* **2019**, *8*, 513. [CrossRef]

22. Devries, R.C. Fault-Tolerant Techniques for Radiation Environments. *IEEE Trans. Nucl. Sci.* **1979**, *26*, 4320–4326. [CrossRef]

23. Na, J.; Lee, D. A study on the reliability improvement factor of fault tolerant mechanisms. In Proceedings of the Safe comp 2013 Fast Abstract, Toulouse, France, 24–27 September 2013.

24. Samudrala, P.K.; Ramos, J.; Katkoori, S. Selective triple modular redundancy for SEU mitigation in FPGAs. *Proc. Mil. Aerosp. Appl. Program. Logic Devices (MAPLD)* **2003**, 344–350. [CrossRef]

25. Morgan, K.; Caffrey, M.; Graham, P.; Johnson, E.; Pratt, B.; Wirthlin, M. SEU-induced persistent error propagation in FPGAs. *IEEE Trans. Nuclear Sci.* **2005**, *52*, 2438–2445. [CrossRef]

26. de Lima Kastensmidt, F.G.; Neuberger, G.; Hentschke, R.F.; Carro, L.; Reis, R. Designing fault-tolerant techniques for SRAM-based FPGAs. *IEEE Design Test Comput.* **2004**, *21*, 552–562. [CrossRef]

27. Lho, Y.H.; Hwang, Y.S.; Lee, S.Y. A study on radiation effects on PWM-IC controller of DC/DC power buck converter. In Proceedings of the 2013 13th International Conference on Control, Automation and Systems (ICCAS 2013), Gwangju, South Korea, 20–23 October 2013; pp. 1754–1757.

28. Adell, P.C.; Schrimpf, R.D.; Holman, W.T.; Boch, J.; Stacey, J.; Ribero, P.; Sternberg, A.; Galloway, K.F. Total-dose and single-event effects in DC/DC converter control circuitry. *IEEE Trans. Nucl. Sci.* **2003**, *50*, 1867–1872. [CrossRef]

29. Baronti, F.; Adell, P.C.; Holman, W.T.; Schrimpf, R.D.; Massengill, L.W.; Witulski, A.; Ceschia, M. DC/DC switching power converter with radiation hardened digital control based on SRAM FPGAs. In Proceedings of the Military and Aerospace Programmable Logic Devices (MAPLD) Workshop, San Diego, CA, USA, 19–22 May 2014.

30. Santos, M.; Ribeiro, H.; Martins, M.; Guilherme, J. Switch mode power supply design constraints for space applications. In Proceedings of the Proc Conf. on Telecommunications–ConfTele, Peniche, Portugal, 9–11 May 2007; pp. 157–160.

31. Rudrakshi, S.; Midasala, V.; Bhavanam, S.N. Implementation of FPGA based fault injection Tool (FITO) for testing fault tolerant designs. *Int. J. Eng. Technol.* **2012**, *4*, 522. [CrossRef]

Article

A Virtual Device for Simulation-Based Fault Injection

Maria Muñoz-Quijada, Luis Sanz and Hipolito Guzman-Miranda *

Department of Electronic Engineering, Universidad de Sevilla, Camino de los Descubrimientos s/n,
41092 Sevilla, Spain; maria_munoz@us.es (M.M.-Q.); luis.sanz@gmail.com (L.S.)
* Correspondence: hguzman@us.es; Tel.: +34-954-481-298

Received: 19 October 2020; Accepted: 22 November 2020; Published: 24 November 2020

Abstract: This paper describes the design and implementation of a virtual device to perform simulation-based fault injection campaigns. The virtual device is fully compatible with the same user software that is already being used to perform fault injection campaigns in existing FPGA (Field Programmable Gate Array)-based hardware devices. Multiple instances of the virtual device can be launched in parallel in order to speed-up the fault injection campaigns, without any preexisting limitations on number, such as available license seats, since the virtual device can be compiled with the open-source simulator GHDL. This virtual device also allows one to find bugs in both software and firmware, and to reproduce in simulation, with total visibility of the internal states, corner cases that may have occurred in the real hardware.

Keywords: fault injection; simulation; VHDL; single event effects; open source tools

---

1. Introduction

1.1. Background

Electronic devices are susceptible to damage due to external ionizing radiation. These effects are traditionally classified between two groups, according to whether the effects emerge from the gradual degradation of the semiconductor properties due to the accumulated effects of multiple particles, or a single ionizing particle impacting a particularly sensitive volume inside the material. This latter category is commonly referred to as Single Event Effects (SEE), and there exist some design hardening strategies designers can follow in order to harden electronic circuits against some of these effects.

SEEs cause anomalies in the behavior of electronic systems [1], which may lead to catastrophic consequences, especially in applications where a high level of reliability and security is required. These effects can be classified as destructive, whether they cause permanent damage to the device, such as Single Event Latchup (SEL) or Single Event Burnout (SEB), or non-destructive, when they only affect the expected behavior of the device without physically destroying it [2], like Single Event Transient (SET) or Single Event Upset (SEU).

1.2. Problem of Interest

In the space sector, which is the most affected by these radiation effects, the use of SRAM (Static Random Access Memory) COTS (Commercial Off-The-Shelf) FPGAs (Field Programmable Gate Arrays) are becoming increasingly important [3,4]. These FPGAs are not necessarily hardened against SEUs so they must be hardened with a mitigation strategy to mitigate these logic effects when deploying such high-performing FPGAs in missions that require high reliability. There exist many techniques that can be used to mitigate these effects such as [5–7]. A compilation of techniques that can be applied in order to mitigate radiation effects, organized by different design stages and abstraction levels, can be found in [8].

The most common mitigation techniques use spatial redundancy, which is also known as hardware redundancy. These techniques involve adding more hardware, thus consuming more resources, in order to detect discrepancies between replicated elements.

Dual Modular Redundancy (DMR) consists of duplicating hardware elements to allow for the detection, but not correction, of faults. Triple Modular Redundancy (TMR) triplicates the hardware elements and inserts a majority voter circuit in order to choose the correct output in case of having a discrepancy between the replicated elements. While these techniques can be applied either to small elements [6] or complete modules, DMR is usually applied to full modules.

Error Detection and Correction (EDAC) algorithms allow one to detect and correct errors in data words without requiring a triplication of the information, and are typically used in memories. For this purpose, Hamming codes are commonly-used error correcting codes.

In order to apply these techniques properly, a key recommendation is to carry out a study in the early phases of a design so as to determine the behavior of the electronic devices against these radiation effects. This can tell the designer which elements of their design are most susceptible of propagating erroneous values to the circuit outputs, producing failures, when being corrupted by these effects. The most sensitive design elements can then be hardened to optimally achieve the design reliability required for a specific mission. In this context, fault injectors are a useful tool to study the behavior of electronic systems against SEE.

1.3. Literature Survey

There are many types of fault injectors that can be found in the literature. They can be classified into five main categories according to the type of injection technique used [9]. These are: Hardware-based fault injection, software-based fault injection, FPGA-based fault injection, simulation-based fault injection, and hybrid fault injection.

1.3.1. Hardware-Based Fault Injection

In the first category, the device under test is physically attacked by external sources, such as a laser [10] or by injecting the current through the pins of the device with active probes, as in [11]. These types of injectors may be destructive to the device causing an over-increase of the initial budget for the project. On the other hand, some of the advantages of these techniques are the wide range of possible locations that can be injected in comparison with other techniques and the accuracy of the results obtained, since real hardware and software are being used.

1.3.2. Software-Based Fault Injection

Software-based fault injection techniques use software to insert the faults in the DUT (Design Under Test) [12,13]. This technique has the benefit of being a portable tool, allowing its use in many platforms without damaging the DUT, but it has two main drawbacks: It only can be used in microprocessor designs, and it cannot access the entire device to insert the faults, only the registers that are available through the microprocessors' ISA (Instruction Set Architecture). Furthermore, the technique is invasive because the software code has to be instrumented in order to inject the faults.

1.3.3. FPGA-Based Fault Injection

Instrumentation techniques can also be applied to HDL code leading to instrumented FPGA-based fault injection. The main criticism that this technique receives is that the circuit that is being tested is not the same as the one that is intended to be deployed on the final mission application, since the VHDL or Verilog code has to be modified in order to perform the fault injection. In critical applications, invasive techniques have the risk of masking functional failures due to the changes added to the DUT. However, these techniques do not require specific hardware nor hidden knowledge of the internal mechanisms of the chosen FPGA, and thus can be applied to many commercial development kits. An example of these techniques can be found in [14].

Some SRAM-based FPGA families include internal circuitry that can be used to read and write internal circuit values, which allows injecting faults in an FPGA design without instrumenting the HDL code. This technique is called non-instrumented FPGA-based fault injection. Traditionally, researchers have developed their own techniques based on limited documentation and reverse engineering in order to inject faults using the internal FPGA circuitry, when the observing and controlling capabilities are implemented in the silicon [15]. The least invasive way of performing this is to have a dedicated chip to perform fault injection, input/output vector control, and campaign execution, leaving the full target FPGA to host the user design [16].

Due to the increase in popularity of the fault injection techniques, nowadays some FPGA vendors are providing IP (Intellectual Property) cores to perform the SEU injection [17,18]. The use of these SEU injection IP cores is less invasive than instrumenting the complete HDL design, but nevertheless requires some changes to the DUT, at least to instantiate the required IP cores and add some kind of control logic to manage the tests. We could call this technique minimally-instrumented fault injection. An example of the application of this technique can be found in [19] where some debugging facilities from Altera FPGAs are used to inject faults in the device under test.

1.3.4. Simulation-Based Fault Injection

Simulation-based injectors have the benefit of being a flexible and inexpensive tool. They use a simulation model of the DUT, which can be described in hardware description language such as VHDL. This technique allows full control of the injection mechanisms as can be seen in [20].

A good review of the different techniques that can be used to perform simulation-based fault injection in VHDL can be found in [21]. According to this reference, there are three possible techniques that can be used:

Simulator commands technique

This is the simulation equivalent to the non-instrumented fault injection technique. When simulator commands can be used to inject the faults, there is no need to instrument the VHDL design, which avoids the aforementioned issues related to design instrumentation. Depending on the fault model used, the required simulator command sequence may vary. The main drawback of this technique is that not all simulators support these commands. A second drawback of this technique is that, depending on how the faults are injected, the technique could be fairly demanding to implement. For example, using interactive commands is fairly easy, but implementing a complete solution that uses the Verilog Procedural Interface of a simulator can be very complex because different simulation objects (such as signals, ports, or variables) may be accessed in different ways [22].

Saboteurs technique

This technique consists of adding VHDL components that modify the characteristics of signals that interconnect VHDL modules of the design under test. This way, values and timing characteristics of these interconnection signals can be altered during the simulation. The main drawback of this technique is that the circuit has to be instrumented, but on the other hand, it can be applied using any VHDL simulator. Since the saboteurs must not interfere with the normal operation of the circuit, a number of control and selection signals must be added to the design and also managed, either through the simulator commands or through extra design inputs.

Mutants technique

The mutants technique is similar to the saboteurs technique in the sense that the VHDL design is instrumented, but in the case of the mutants, design components are replaced by mutant components. These mutants operate like the original component in the absence of faults, but one or more parts of its functionality are altered when activated. The VHDL configuration keyword allows one to select, for each component, either its original architecture or one of a set of mutant architectures. In order to change the configuration of a component, the architecture to component binding (meaning which architecture a specific component will have) and the new configuration must be recompiled, but since this is a partial compilation, there is no need for recompiling the complete design. Since this technique does not add new components and instead just changes the architecture of the already existing design components, in the absence of mutations the obtained design is equal to the original design.

While every researcher or engineer might have their own preference, it must be noted that these techniques are in no way exclusive as more than one of them could be applied at the same time, for example including both mutants and saboteurs in the same instrumented design.

1.3.5. Hybrid Fault Injection

The last category of injectors use a combination of the aforementioned techniques to improve the injection capabilities in conjunction [23].

1.4. Scope and Contribution of This Paper

The paper presents a virtual device to perform simulation-based fault injection using open source tools. The virtual device is fully compatible with the software used for an existing FPGA-based fault injection platform and it also allows one to verify both software and firmware parts of this fault injection platform. As mentioned in the previous section, simulation-based injectors have the advantage of being inexpensive and portable tools unlike the other techniques exposed above.

One of the differences between simulation-based techniques mentioned like [20] and the proposed approach is the use of open source tools, which allows more flexibility in the use and applications of the proposed approach. The proposed approach uses the GHDL and cocotb tools to simulate the design and perform the fault injection, thus allowing the user free and complete use of the capabilities of the tools, for example running multiple instances of the device without any licensing limitations to perform multiple injections in parallel.

Another contribution of the proposed approach is that, although the approach can be slower than other alternatives based on simulation, since the test shell that goes in the service FPGA is also simulated and in return it is guaranteed that it is fully compatible with the software that manages the real hardware. This allows the tool to be used as a debugger for the firmware that goes in the actual hardware of the FPGA-based fault injection platform. Furthermore, this compatibility allows one to combine any number of physical and virtual devices in order to accelerate the execution of the fault injection campaigns.

1.5. Organization of the Paper

The paper is structured as follows: Section 2 describes the architecture of the proposed approach. In Section 3, experimental results of the fault injection campaigns are obtained using the virtual device, and the advantages and disadvantages of the approach are discussed. Finally, the conclusions and future work are presented in Section 4.

2. Virtual Device Architecture

The virtual device (also shortened as vdev) is a VHDL model of the firmware architecture for an FPGA-based fault injection platform known as FTU-VEGAS. This platform is developed by Universidad de Sevilla through the European H2020 project VEGAS, (Validation of high capacity rad-hard FPGA and software tools). The system has two FPGAs, one to perform the injections, manage the command set, and compare faulty outputs with the golden outputs (outputs without injections), and another FPGA which hosts the Design Under Test (DUT). The software part of the system is named tntsh (Test aNalysis Tools shell). This software operates the hardware by sending the necessary commands and data to perform the injection campaigns, and also receives and stores the results. The architecture for the vdev proposed is the one shown in Figure 1. The vdev communicates with the same software (tntsh) used for the physical hardware through a pair of pipes. The virtual device follows a modular architecture where each module communicates with another through a pair of streams. The functionality and architecture of each module are presented in the following subsections, starting from the top level.

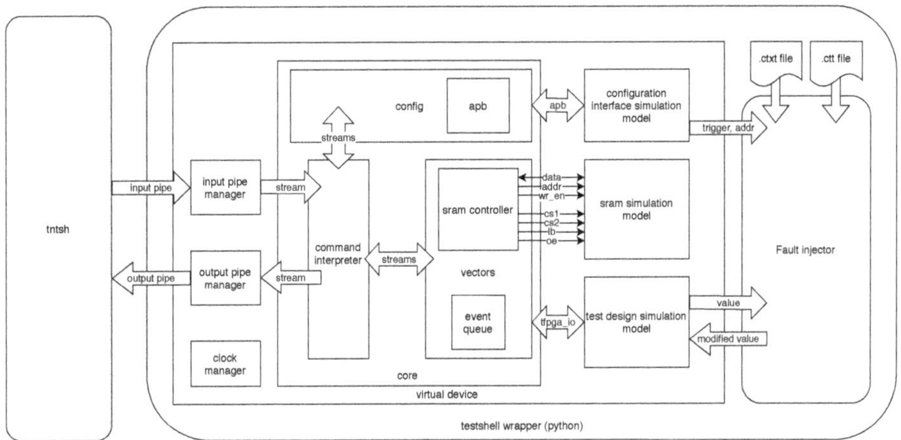

Figure 1. Firmware architecture of the virtual device and communication with tntsh.

2.1. Top Level Module

This is the top level of the virtual device. This module receives the commands from tntsh and returns the requested data to it through a pair of input/output pipes.

2.2. SRAM Simulation Model

This is a model for the R1WV6416R SRAM device from RENESAS, used to store the input/output vectors of a fault injection campaign in an internal format called wave. The wave contains the bit array that contains the concatenated inputs for the DUT each clock cycle, and the corresponding bit array with the concatenated outputs, and is stored in the SRAM using a simple compression schema.

2.3. Test Design Simulation Model

Instances the design under test.

2.4. Core Module

This module is responsible for accepting instructions from the tntsh software and returning the appropriate values. It instances the command interpreter, which manages the commands received from the software and also interfaces with the vectors and configuration modules. The data interchange between the command interpreter and the rest of the modules is managed by stream modules.

2.5. Stream Module

The stream is the interface used for interchanging data between modules. It is composed of an encoder, a FIFO (First In First Out) memory, and a decoder. Every main module inside the virtual device uses an input stream to request input data and an output stream to provide output data. The stream has been designed to simplify the exchange of multiple data of different widths between modules. Internal data inside the virtual device may have different widths, for example, an injection address may be a 32-bit value but a time value in cycles may be a 64-bit value, while a command always has an 8-bit value. In addition, in order to save SRAM memory space, the width of input and output vectors depends on the characteristics of each design under test. This module needs thus to manage a stream of data of different sizes, with a size between 1 and 8 bytes for each data.

Figure 2 shows the architecture of the stream. The process for module A to send a single multi-byte data to module B is as follows:

- A writes in the stream:

  1. A waits for the write side of the stream to be ready (*wr_ready* active). If the write side is not ready, *wr_op* must be set to zero;
  2. A sets *wr_op* to the number of bytes that need to be written (*N*), while at the same time sets *wr_data*(N*8-1 *downto 0)* to the value of the data word to write.

- B reads from the stream:

  1. B waits for the read side of the stream to be ready (*rd_ready* active). If the read side is not ready, *rd_op* must be set to zero;
  2. B sets *rd_op* to the number of bytes that need to be read (*N*);
  3. Starting from the next clock cycle, when *rd_ready* is asserted again, *rd_data* is valid, from which B reads the least significant *N* bytes.

It must be noted that B can ask for data before A writes anything into the stream, and this will not cause any issue, since the stream will not assert *rd_ready* until it has enough data, which effectively waits for A to write the requested data.

When bidirectional communications are needed, for example in case of a module sending commands to a submodule and reading the responses to these commands, two streams can be instanced, one for each data direction.

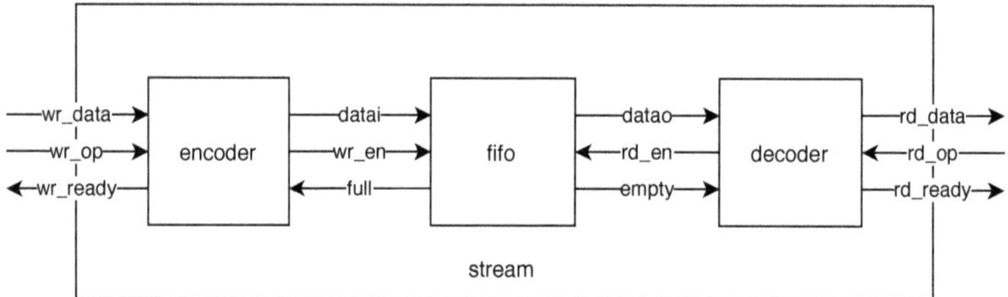

Figure 2. Stream module. For simplicity, global clock and reset connections are not shown. A dual-clock fifo may be used to decouple clock domains and in that case, the encoder should use the same clock as the write side of the fifo, and the decoder should use the same clock as the read side of the fifo.

2.6. Config Module

The configuration module is responsible for reading and writing in the simulated target FPGA. It can configure a valid bitstream, and perform I/O operations on individual configuration bits, such as bit flips. It interfaces with a model of the target FPGA (configuration interface simulation model) which interchanges data with the logic of the configuration module using the APB (Advanced Peripheral Bus) protocol.

2.7. Vectors Module

The module is responsible for handling the input and output vectors, both the golden data and the experiment results, as well as handling the emulation clock. It includes the SRAM controller to store and manage the input/output vectors in the SRAM and event queue.

The Event Queue

Every time a special condition is detected during the fault injection (such as discrepancy with golden outputs or end of vectors), an event is raised. When an event is raised, a new item is added to the event queue with each item containing two data: A 1-byte mask of all the events raised and an 8-byte value containing the cycle in which these events were raised.

The event queue allows one to configure flags for the fault injection campaign that can alter the course of the test depending on what happens during the experiment, without continuously communicating with the software. For example, by stopping a run (a complete execution of the test vectors with zero, one or more injections) after detecting damage, without simulating the rest of the clock cycles. The complete event queue can be read by the software after a single run, reducing communication overhead.

2.8. Flow Process and Required Files

The process of injecting faults is made using cocotb [24]. Cocotb is a cosimulation testbench environment written in python. It is an open-source tool which can be used in multiple operating systems. The GHDL open-source simulator is also used to compile the source VHDL code of the vdev to obtain the executable file. Cocotb accesses the values inside the simulation using the simulator's VPI (Verilog Procedural Interface). A testshell wrapper for the GHDL executable has been written in python so cocotb can be used to access the internal simulation values. With respect to the classification described in Section 1.3.4, we can consider this technique inside the 'simulator commands' category.

The tntsh software can be used interactively since it provides a TCL (Tool Command Language) shell, but it can also be used in batch mode by providing .tcl scripts with the commands to execute. Makefiles can be used then to automate the execution of multiple fault injection campaigns, using one or multiple instances of the vdev.

The following files are needed to perform a fault injection campaign:

- pin file:
  Contains the inputs, outputs, and clock pin signal of the DUT. This file must be written by the user, using a very simple format to indicate signal names, directions, and widths;
- nxb file:
  The configuration bitstream of the DUT. Generated by the NXmap FPGA vendor tool. While this file is obviously required when using the real hardware, a dummy nxb can be used when injecting faults with the vdev;
- vcd file:
  A value change dump with the recorded input/output vectors obtained by simulating the design. The user must generate this file using their own testbench with any simulator that supports the generation of VCD files;
- ctxt file:
  A logic location file that shows the position of the user logic inside the bitstream. Generated by the NXmap tool. This file can also be substituted for a dummy file when using the vdev, removing the need for the proprietary NXmap tool;
- ctt file:
  This file relates the register names inside the ctxt file with the hierarchical signal names inside the vdev. This file is generated semi-automatically by processing the ctxt file;
- test.tcl:
  A tcl script with the commands to be executed to perform a fault injection. It loads the configuration, the vectors, and the location files and selects the options desired to perform the campaign. These are the same commands that are supported by the real hardware. The tcl file is not strictly necessary, since the commands can be entered interactively in the tntsh shell.

It must be noted that the nxb and ctxt files are equivalent to bitstream and logic location files generated by software from other vendors, such as Xilinx. The specific nxb and ctxt files are used here so both the software and virtual device remain compatible with the hardware of the fault injection platform, but other file formats could be supported.

The tntsh and virtual device support multiple injection campaign options that can be selected to customize the fault injection experiments. The injector function and injection mask can be also selected. These options are listed below:

- Campaign options:

    - check_residual_damage:
      Checks for damage in the output vectors, during the cycles before injection, for every run;
    - damage_per_run:
      Maximum number of output damages logged per run;
    - drop_on_damage:
      Stop runs after damage_per_run damages;
    - export_io:
      Read the complete faulty vectors, in wave format, after each run;
    - unflip_after_run:
      Attempts to fix the damage to the emulated circuit by unflipping the bits associated with the registers that are changed, at the end of the run;
    - blocking:
      Run campaigns on the foreground, instead of as a background process;
    - workdir:
      Log campaigns in a custom path.

- Injectors:

    - inj_gauss:
      The number of injections is determined randomly according to the normal distribution;
    - inj_binomial:
      The number of injections is determined randomly according to the binomial distribution;
    - inj_poisson:
      The number of injections is determined randomly according to the Poisson distribution;
    - inj_exhaustive:
      This injector will generate one injection per run in every possible combination of the target cycle and register lists, then interrupt the campaign;
    - inj_file:
      This injector reads its injections from a string formatted as a csv file;
    - inj_clean
      An injector that generates empty injection lists.

- Injection Masks:
  The injection masks eliminate injections from the ones generated by the injector according to some criteria.

    - mask_bernoulli:
      Set up a mask that discards injections with a threshold probability.;
    - mask_none:
      Unset current injection mask, and all injections will be used during the campaign.

2.9. Technical Requirements for the Virtual Device

While there are no specific hardware requirements for the use of the virtual device, it needs the following software:

- The GHDL simulator, which is available both for Linux and Microsoft Windows;
- A mechanism to create unix pipes or pipes that behave as such;
- The tntsh software, which in turn requires a C compiler that supports the C++17 revision of the standard for the C++ programming language (such as gcc or clang), and some libraries readily available in most modern GNU/Linux systems;

- The cocotb coroutine simulation framework, which in turn requires python 3.

The experimental results for this paper have been obtained in an Acer EX2540 series computer, model NX.EFHEB.2002, with 8 GB of RAM and an Intel Core i5 7200U processor, running the Debian GNU/Linux Operating System, version 10.

3. Experimental Results

A number of DUTs of increasing complexity have been chosen to perform fault injection with the vdev. The selected designs are described below:

- counter:
  Implements an 8-bit counter;
- adder acum:
  This design accumulates the value of an 8-bit input vector in a 20-bit vector;
- shiftreg:
  Implements an 8-bit shift register;
- b13:
  An interface to meteo sensors [25];
- FIFO:
  A simple 32-bit FIFO memory [26];
- pcm:
  An Integrated Interchip Sound (IIS) interface for the PCM3168 codec [27].

For each DUT, a campaign using a gaussian injector and no injection masks was performed. The gaussian injector was configured with $\mu = 1$ and $\sigma = 0$, so a single SEU was injected each run. Each campaign performed a total of 1000 injections among the list of candidates, which may or may not have propagated to the primary outputs, causing output damage. For each design, the Architectural Vulnerability Factor (AVF), which is the percentage of damages obtained over the total of injections performed [28], was calculated and is presented in Tables 1–6. Note that registers that do not produce output errors when injected are not shown in the tables, but are included in the global AVF calculations of the design. Sometimes designs can break when adapting them to a fault injection platform if the process is not made with special care. To check the correct functionality of the designs before performing a campaign, an emuvssim test was performed for each design. Emuvssim tests check the emulation in the virtual device and simulation with a test bench given by the user match, by comparing both waveforms. This allows one to prove that the virtual device does not break functionality and also helps to verify the firmware/software of the fault injection platform. All designs passed the emuvvsim tests.

Table 1. Campaign for counter design.

Design: Counter					
reg_name	bits	damages_per_reg	injections_without_damage	injections_per_reg	avf (%)
add_L26_stage1:SUM1	1	120	0	120	100
add_L26_stage1:SUM2	1	121	4	125	96.8
add_L26_stage1:SUM3	1	117	2	119	98.3
add_L26_stage1:SUM4	1	127	3	130	97.7
add_L26_stage2:SUM1	1	148	2	150	98.7
add_L26_stage2:SUM2	1	118	4	122	96.7
add_L26_stage2:SUM3	1	95	0	95	100
add_L26_stage2:SUM4	1	139	0	139	100
TOTAL	8	985	15	1000	98.5

Table 2. Campaign for adder_acum design.

Design: Adderacum					
reg_name	bits	damages_per_reg	injections_without_damage	injections_per_reg	avf (%)
add_L23_stage1:SUM1	1	39	3	42	92.9
add_L23_stage1:SUM2	1	43	2	45	95.5
add_L23_stage1:SUM3	1	51	1	52	98
add_L23_stage1:SUM4	1	36	0	36	100
add_L23_stage2:SUM1	1	44	0	44	100
add_L23_stage2:SUM2	1	42	2	44	95.5
add_L23_stage2:SUM3	1	46	3	49	93.9
add_L23_stage2:SUM4	1	60	4	64	93.8
add_L23_stage3:SUM1	1	41	5	46	89.1
add_L23_stage3:SUM2	1	39	1	40	97.5
add_L23_stage3:SUM3	1	50	1	51	98
add_L23_stage3:SUM4	1	47	2	49	95.9
add_L23_stage4:SUM1	1	38	1	39	97.4
add_L23_stage4:SUM2	1	47	2	49	95.9
add_L23_stage4:SUM3	1	58	4	62	93.5
add_L23_stage4:SUM4	1	48	0	48	100
add_L23_stage5:SUM1	1	36	3	39	92.3
add_L23_stage5:SUM2	1	58	1	59	98.3
add_L23_stage5:SUM3	1	45	3	48	93.8
add_L23_stage5:SUM4	1	37	2	39	94.9
TOTAL	21	95	905	1000	90.5

Table 3. Campaign for shiftreg design.

Design: Shiftreg					
reg_name	bits	damages_per_reg	injections_without_damage	injections_per_reg	avf (%)
reg_reg	8	100	900	1000	10
TOTAL	8	100	900	1000	10

Table 4. Campaign for B13 design.

Design: B13					
reg_name	bits	damages_per_reg	injections_without_damage	injections_per_reg	avf (%)
rdy_reg	1	22	4	26	84.6
send_en_reg	1	17	4	21	80.9
tre_reg	1	1	16	17	5.9
TOTAL	49	40	960	1000	4.0

Table 5. Campaign for FIFO (First In First Out) design.

Design: FIFO					
reg_name	bits	damages_per_reg	injections_without_damage	injections_per_reg	avf (%)
empty_reg	1	12	39	51	23.5
head_reg	8	91	328	419	21.7
looped_reg	1	8	44	52	15.4
tail_reg	8	104	337	441	23.6
full_reg	1	12	25	37	32.4
TOTAL	19	227	773	1000	22.7

Table 6. Campaign for pcm3168 design.

Design: pcm3168					
reg_name	bits	damages_per_reg	injections_without_damage	injections_per_reg	avf (%)
I2S_IN_1 I DATA_L_reg	23	176	152	328	53.7
I2S_IN_1 I s_current_lr_reg	1	2	12	14	14.3
I2S_IN_1 I shift_reg_reg	23	110	212	322	34.1
TOTAL	74	288	712	1000	28.8

3.1. Campaign Execution Times

To save execution time, multiple instances of the virtual device can be launched in parallel. The maximum number of instances depends on the processor capacity of the user thus allowing a faster performance than when using proprietary tools, if enough processing capability is available.

Table 7 shows the percentage of execution speed improvement when using two, four, and eight devices:

Table 7. Execution time improvement.

Design	2 Devices (%)	4 Devices (%)	8 Devices (%)
counter	40.83	64.20	74.35
adder_acum	45.71	62.64	71.81
dualcounter	42.48	65.08	73.77
shiftreg	42.98	67.27	76.83
b13	49.10	66.42	77.64
fifo	47.53	67.69	76.12
pcm	48.48	67.50	77.40

3.2. Discussion

The main disadvantage of the proposed approach is that it is slower than FPGA-based fault injection. Conversely, one of its advantages is that it does not require any hardware devices.

An advantage that can mitigate the previous disadvantage of this approach is that the fault injection campaigns can be parallelized by instantiating multiple virtual devices, up to the available computing capacity. By using open source tools, this approach does not have any arbitrary license-based limitations to the number of devices that can be executed at the same time.

Another advantage of the approach is the compatibility with the same software that is used with the real hardware. The cost of having this compatibility is the increase of complexity in the virtual device.

The current version of the virtual device requires recompiling the VHDL source when new designs are added. In order to make the device available to more users and simplify the design preparation, it would be a good idea to separate the two elements. For example, an object file for the virtual device could be provided that the user could link to the object files of their design under test.

Another issue to consider is that special care must be taken in the timing of the forcing and release actions of the signal where the fault is being injected. If the signal force is released before the active clock cycle of the design under test, it is possible that the fault is erased and not captured by the design. However, if the signal force is released after the active clock cycle of the design under test, there is the risk that the fault does not correctly propagate (for example, in case of the fault crossing some logic cones and propagating to the input of the same flip-flop) or that it remains for more time than is needed, which would result in an incorrect SEU model. This was solved by using the trigger methods provided by cocotb to synchronize with respect to the clock signal for the design under test.

A current limitation of this approach is that the latest version of cocotb (1.5.dev0) cannot access signals inside a record with the latest version of GHDL (1.0-dev/v0.37.0), which may impact the fault coverage of complex designs that use these datatypes. This is a known limitation of the simulator and is currently an open issue in the GHDL issue tracker, so it can be expected to be fixed in the future. A possible workaround until this limitation is removed, is to inject the faults in a post-synthesis version of the design under test. To achieve this, the synthesis of the design under test must be performed, and afterwards a netlist of the synthesized design must be generated in VHDL format. For example, the *netgen* tool from Xilinx allows to do this.

4. Conclusions and Future Work

A virtual device to perform fault injection by simulation was designed, developed, and demonstrated. This virtual device is also a model of the FTU-VEGAS fault injector by emulation firmware, extended with fault injection capabilities, and is fully compatible with the software that communicates with real hardware. A set of injection campaigns with increasing complexity was performed and the results of the Architectural Vulnerability Factor for these campaigns were exposed. The virtual device demonstrated to not disturb the correct functionality of the test designs in the absence of injected faults, with the emulation versus simulation tests. The virtual device can be fully compiled and used with only free and open source software, avoiding the use of expensive proprietary simulators, and campaign speed can be parallelized by running multiple instances of the virtual device without any restrictions, which helps to bridge the speed gap with respect to FPGA-based solutions, when running the tests in powerful servers with multiple processor cores.

Future work will include comparing the results obtained with the virtual device with the results obtained using real hardware when it becomes available and decoupling the HDL compilation of the virtual device and the DUTs, so binaries of the virtual device can be distributed and linked to or co-simulated with a third party's confidential designs.

Author Contributions: Conceptualization, H.G.-M.; Methodology, H.G.-M. and M.M.-Q.; Validation and bug fixing: M.M.-Q., L.S. and H.G.-M.; Software, L.S.; Simulation, M.M.-Q.; Fault injection, M.M.-Q.; Analysis, M.M.-Q. and H.G.-M.; Investigation, M.M.-Q. and H.G.-M.; Writing—original draft preparation, M.M.-Q. and H.G.-M.; Writing—review and editing, H.G.-M. and M.M.-Q.; Supervision, H.G.-M.; Project administration, H.G.-M. All authors have read and agreed to the published version of the manuscript.

Funding: This work was supported by the European Commission, through the project "VEGAS: Validation of European high capacity rad-hard FPGA and software tools", project ID 687220.

Acknowledgments: The authors would like to thank the authors and maintainers of the free and open source tools used in this paper: Tristan Gingold and the rest of GHDL contributors, Potential Ventures, and the rest of cocotb developers and maintainers (especially Luke Darnell for adding the signal force and release capabilities and Stefan Biereigel for porting them to the latest master version of cocotb), the GNU project, and the Python software foundation. The authors also want to thank NanoXplore for providing a version of the NXmap software.

Conflicts of Interest: The authors declare no conflict of interest.

Abbreviations

The following abbreviations are used in this manuscript:

APB	Advanced Peripheral Bus
AVF	Architectural Vulnerability Factor
BRAM	Block Random Access Memory
COTS	Commercial Off-The-Shelf
CPU	Central Processing Unit
DUT	Device Under Test
FIFO	First In First Out
FPGA	Field Programmable Gate Array
FSM	Finite State Machine
FT-Unshades	Fault Tolerance - Universidad de Sevilla Hardware Debugging System
FTU-VEGAS	FT-Unshades for the Validation of European high capacity rad-hard FPGA and Software tools
FW	Firmware
HDL	Hardware Description Language

HW	Hardware
IP	Intellectual Property
OS	Operating System
SEB	Single Event Burnout
SEE	Single Event Effect
SEL	Single Event Latchup
SET	Single Event Transient
SEU	Single Event Upset
SRAM	Static Random Access Memory
SW	Software
VCD	Value Change Dump
VDEV	Virtual DEVice
VHDL	Very High Speed Integrated Circuit Hardware Description Language

References

1. Duzellier, S. Radiation effects on electronic devices in space, Radiation effects on electronic devices in space. *Aerosp. Sci. Technol.* **2005**, *9*, 93–99. [CrossRef]
2. Dodd, P.E.; Massengill, L.W. Basic mechanisms and modeling of single-event upset in digital microelectronics. *IEEE Trans. Nucl. Sci.* **2003**, *50*, 583–602. [CrossRef]
3. Bedi, R. First Design-In Experiences of Xilinx's, 20 nm, Kintex Ultrascale KU060 for Space Applications and 16 nm UltraScale+ RFSoC for Ground Segment. In *2018 SpacE FPGA Users Workshop*, 4th ed.; European Space Research and Technology Centre (ESTEC), European Space Agency (ESA): Noordwijk, The Netherlands, 2018. [CrossRef]
4. Michel, H.; Guzmán-Miranda, H.; Dörflinger, A.; Michalik, H.; Echanove, M.A. SEU fault classification by fault injection for an FPGA in the space instrument SOPHI. In Proceedings of the 2017 NASA/ESA Conference on Adaptive Hardware and Systems (AHS), Pasadena, CA, USA, 24–27 July 2017; pp. 9–15. [CrossRef]
5. Larchev, G.V.; Lohn, J.D. Evolutionary based techniques for fault tolerant field programmable gate arrays. In Proceedings of the 2nd IEEE International Conference on Space Mission Challenges for Information Technology (SMC-IT'06), Pasadena, CA, USA, 17–20 July 2006. [CrossRef]
6. Muñoz-Quijada, M.; Sanchez-Barea, S.; Vela-Calderon, D.; Guzman-Miranda, H. Fine-Grain Circuit Hardening Through VHDL Datatype Substitution. *Electronics* **2019**, *8*, 24. [CrossRef]
7. Kastensmidt, F.L.; Carro, L.; Reis, R. *Fault-Tolerance Techniques for SRAM-Based FPGAs*; Springer: Boston, MA, USA, 2006. [CrossRef]
8. ECSS Secretariat, ESA-ESTEC, Requirements & Standards Division, "ECSS-Q-HB-60-02A: Space product assurance. Techniques for Radiation Effects Mitigation in ASICs and FPGAs Handbook"; ESA Requirements and Standards Division, Noordwijk, The Netherlands. 2016. Available online: https://escies.org/download/webDocumentFile?id=64426 (accessed on 23 October 2020).
9. Ziade, H.; Ayoubi, R.A.; Velazco, R. A Survey on Fault Injection Techniques. *Int. Arab J. Inf. Technol.* **2004**, *1*, 171–186.
10. Courbon, F.; Fournier, J.J.; Loubet-Moundi, P.; Tria, A.; Combining Image Processing and Laser Fault Injections for Characterizing a Hardware AES. *IEEE Trans. Comput. Aided Des. Integr. Circuits Syst.* **2015**, *34*, 928–936. [CrossRef]
11. Arlat, J.; Crouzet, Y.; Laprie, J.C. Fault injection for dependability validation of fault-tolerant computing systems. In Proceedings of the 19th International Symposium on Fault-Tolerant Computing, Digest of Papers, Chicago, IL, USA, 21–23 June 1989; pp. 348–355. [CrossRef]
12. Mansour, W.; Velazco, R.; El Falou, W.; Ziade, H.; Ayoubi, R. SEU simulation by fault injection in PSoC device: Preliminary results. In Proceedings of the 2012 2nd International Conference on Advances in Computational Tools for Engineering Applications (ACTEA), Beirut, Lebanon, 12–15 December 2012; pp. 330–333. [CrossRef]
13. Da Silva, A.; Gonzalez-Calero, A.; Martinez, J.F.; Lopez, L.; Garcia, A.B.; Hernández, V. Design and Implementation of a Java Fault Injector for Exhaustif® SWIFI Tool. In Proceedings of the 2009 Fourth International Conference on Dependability of Computer Systems, Brunow, Poland, 30 June–2 July 2009; pp. 77–83. [CrossRef]

14. Civera, P.; Macchiarulo, L.; Rebaudengo, M.; Reorda, M.S.; Violante, M. Exploiting circuit emulation for fast hardness evaluation. *IEEE Trans. Nucl. Sci.* **2001**, *48*, 2210–2216. [CrossRef]
15. Aguirre, M.A.; Tombs, J.N.; Torralba, A.; Franquelo, L.G. UNSHADES-1: An advanced tool for In-System Run-Time Hardware Debugging. In Proceedings of the Field Programmable Logic and Applications, Lisbon, Portugal, 1–3 September 2003; pp. 1170–1173. [CrossRef]
16. Mogollon, J.M.; Guzman-Miranda, H.; Napoles, J.; Barrientos, J.; Aguirre, M.A. FTUNSHADES2: A novel platform for early evaluation of robustness against SEE. In Proceedings of the 2011 12th European Conference on Radiation and Its Effects on Components and Systems, Sevilla, Spain, 19–23 September 2011; pp. 169–174. [CrossRef]
17. Intel. Fault Injection Intel FPGA IP Core User Guide. Available online: https://www.intel.com/content/dam/www/programmable/us/en/pdfs/literature/ug/ug_fault_injection.pdf (accessed on 19 October 2020).
18. Xilinx. Soft Error Mitigation (SEM) Core. Available online: https://www.xilinx.com/products/intellectual-property/sem.html (accessed on 19 October 2020).
19. Mohammadi, A.; Ebrahimi, M.; Ejlali, A.; Miremadi, S.G. SCFIT: A FPGA-based fault injection technique for SEU fault model. In Proceedings of the 2012 Design, Automation & Test in Europe Conference & Exhibition (DATE), Dresden, Germany, 12–16 March 2012; pp. 586–589. [CrossRef]
20. Baraza, J.C.; Gracia, J.; Gil, D.; Gil, P.J. A prototype of a VHDL-based fault injection tool: Description and application. *J. Syst. Archit.* **2002**, *47*, 847–867. [CrossRef]
21. Baraza, J.C.; Gracia, J.; Gil, D.; Gil, P.J. VHDL Simulation-Based Fault Injection Techniques. In *Fault Injection Techniques and Tools for Embedded Systems Reliability Evaluation*; Benso, A., Prinetto, P., Eds.; Frontiers in Electronic Testing; Springer: Boston, MA, USA, 2002; Volume 23. [CrossRef]
22. Sutherland, S. *The Verilog PLI Handbook: A User's Guide and Comprehensive Reference on the Verilog Programming Language Interface*; The International Series in Engineering and Computer Science; Springer: Boston, MA, USA, 2002; Volume 666. [CrossRef]
23. Guthoff, J.; Sieh, V. Combining software-implemented and simulation-based fault injection into a single fault injection method. In Proceedings of the 25th International Symposium on Fault-Tolerant Computing, Digest of Papers, Pasadena, CA, USA, 27–30 June 1995. [CrossRef]
24. Various. Cocotb Documentation. Available online: https://docs.cocotb.org/en/stable/ (accessed on 19 October 2020).
25. Corno, F.; Reorda, M.S.; Squillero, G. RT-level ITC'99 benchmarks and first ATPG results. *IEEE Des. Test Comput.* **2000**, *17*, 44–53. [CrossRef]
26. VHDL Standard FIFO. Available online: http://www.deathbylogic.com/2013/07/vhdl-standard-fifo/ (accessed on 19 October 2020).
27. I^{2}S Interface Designed for the PCM3168 Audio Interface from Texas Instruments. Available online: https://github.com/wklimann/PCM3168 (accessed on 19 October 2020).
28. Mukherjee, S.S.; Weaver, C.T.; Emer, J.; Reinhardt, S.K.; Austin, T. Measuring architectural vulnerability factors. *IEEE Micro* **2003**, *23*, 70–75. [CrossRef]

Publisher's Note: MDPI stays neutral with regard to jurisdictional claims in published maps and institutional affiliations.

 **_electronics_**

Article

Comparison of the Total Ionizing Dose Sensitivity of a System in Package Point of Load Converter Using Both Component- and System-Level Test Approaches

Tomasz Rajkowski [1,2,*], **Frédéric Saigné** [2], **Kimmo Niskanen** [2], **Jérôme Boch** [2], **Tadec Maraine** [2], **Pierre Kohler** [1], **Patrick Dubus** [1], **Antoine Touboul** [2] **and Pierre-Xiao Wang** [1]

[1] 3D PLUS, 408 rue Hélène Boucher, 78532 Buc, France; pkohler@3d-plus.com (P.K.); patrick.dubus@powerlogy.fr (P.D.); pwang@3d-plus.com (P.-X.W.)

[2] Institut d'Electronique et des Systèmes, Université de Montpellier, 860 rue Saint Priest, Bâtiment 5, CC 05001, CEDEX 5, 34095 Montpellier, France; frederic.saigne@ies.univ-montp2.fr (F.S.); kimmo.h.niskanen@jyu.fi (K.N.); jerome.boch@ies.univ-montp2.fr (J.B.); tadec.maraine@ies.univ-montp2.fr (T.M.); antoine.touboul@ies.univ-montp2.fr (A.T.)

* Correspondence: trajkowski.openaccess@gmail.com

Abstract: Testing at system level is evaluated by measuring the sensitivity of point-of-load (PoL) converter parameters, submitted to total ionizing dose (TID) irradiations, at both system and component levels. Testing at system level shows that the complete system can be fully functional at the TID level more than two times higher than the qualification level obtained using a standard-based component-level approach. Analysis of the failure processes shows that the TID tolerance during testing at system level is increased due to internal compensation in the system. Finally, advantages and shortcomings of the testing at system level are discussed.

Keywords: total ionizing dose; system-level testing; point-of-load converter; radiation hardness assurance; system qualification

Citation: Rajkowski, T.; Saigné, F.; Niskanen, K.; Boch, J.; Maraine, T.; Kohler, P.; Dubus, P.; Touboul, A.; Wang, P.-X. Comparison of the Total Ionizing Dose Sensitivity of a System in Package Point of Load Converter Using Both Component- and System-Level Test Approaches. *Electronics* **2021**, *10*, 1235. https://doi.org/10.3390/electronics10111235

Academic Editor: Elias Stathatos

Received: 15 April 2021
Accepted: 19 May 2021
Published: 22 May 2021

Publisher's Note: MDPI stays neutral with regard to jurisdictional claims in published maps and institutional affiliations.

1. Introduction

The qualification of components that will be used for space missions requires test standards that allow the selection of components that will make up the on-board systems [1–3]. With the increasing use of commercial off-the-shelf (COTS) devices or the need to set up selection methodologies for new fields, such as nuclear decommissioning or NewSpace, the question of testing no longer at the component level but at the level of a system arises [4–12]. It is expected that testing at system level may lead to the reduction of the test effort when compared to testing at the component level of all the parts constituting the system—therefore, it may also reduce time-to-market for new products. System-level testing may also provide extended radiation data for some products, e.g., devices that were too difficult to characterize fully with component-level testing. However, currently there is no standard that would describe the qualification process of electronics through radiation testing at system level and this paper aims to evaluate some of the capabilities of this approach.

In this paper, TID system-level tests are discussed. They are typically easier and cheaper to perform than single event effect (SEE) system-level tests, due to relatively high accessibility of the standard test source, the Co^{60} isotope that emits highly penetrating gamma rays, which can penetrate whole electronic boards and systems. The test setup is usually also simpler for TID testing because generally there is no need to monitor the performance of the system under test (SUT) during irradiation; characterization might be performed when irradiation is stopped and the SUT removed from the test area (remote testing), although sometimes it is chosen to perform in situ or even in-flux measurements with automated test equipment [6,7].

Fernández [6] presented a method of TID testing at system (equipment) level for the ITER experiment, where component test standards are followed where applicable (Test Method 1019 from MIL-STD-750E, Test Method 1019 from MIL-STD-883G and ESCC 22900), due to lack of a standard for system-level testing. Because the SUT has both CMOS and bipolar parts, low dose rate irradiation is used as the worst case for this scenario. Irradiation is followed by annealing. The target environment for the system is not very harsh and the predicted dose level to be achieved during operation in that environment is 100 rad. Therefore, it was possible to perform irradiation with a large margin (300 times) by testing the system up to 30 krad in a reasonable time of around 120 h. Such a large margin is favorable for the critical equipment, in the situation when the system-level test is used experimentally for the system qualification and, e.g., a lot-to-lot variability in the system is not known. The system was performing typical functions during irradiation ("test as you fly" approach) and its performance was continuously monitored: an in-flux functional test was performed every 30 min. Only minor functionality errors were observed at the end of the test, but due to the SUT complexity, it was not possible to track which exact component (or group of components) was responsible for this failure (particularly because high-level function health was monitored and not parameters (e.g., electrical parameters) of specific components) [6].

Rousselet [7] described board-level TID tests of the COTS single-board computer (SBC) equipped with an ARM Cortex A8 processor, memories (DDR3, Flash, EEPROM) and voltage regulators. In this test it was possible to observe degradation of specific components or subsystems and to identify the most sensitive components, as well as to define the failure mechanisms—board-level testing enabled better observability than equipment-level testing described in [6], thus providing more data for analysis. Testing of four SBC boards was presented in [7], with three biased boards and one unbiased board. Different functional and electrical parameters were measured (flash memory readout data rate, DDR3 memory input voltage, MicroProcessor Unit voltage, total current) and the measurement spread (between specific SBCs) was observed for these parameters. The test provides information on the dose level for which the SBC (or some of its functions) fail, but the maximum dose level for which it remains fully (or sufficiently) functional is not proposed (as a consequence, margins for such a value are not discussed). It is highlighted that system should be tested in the exact internal configuration in which it will be used in the target mission; differences between biased and unbiased boards were observed. On the other hand, testing in different environmental conditions is also proposed, i.e., in low and high temperatures to assess the worst-case condition for the system [7].

In order to further evaluate the capabilities of the testing at system level, in this paper we have performed TID experiments on a point-of-load (PoL) converter manufactured by 3D-Plus. This point-of-load converter has already been qualified using the ESA standard ECSS-22900 [13]. In addition, all the devices of the complete system have been tested individually following the standards. The PoL converter has also been qualified for a total dose of 50 krad(Si). In this work we have performed irradiations of the complete system to evaluate the total dose that could be reached before failure or being out of specifications and compared it with the 50 krad obtained using a qualification at component level. We have shown that the complete system is still functional after a Co^{60} irradiation at 118 krad(Si). Some parameters have drifted but the system is largely within the specifications. In order to reach the failure of the device, X-ray irradiations have then been performed. A total dose above 400 krad(Si) was necessary to show failures of the tested systems. An analysis has been conducted to find the process at play leading to the degradation of the system. We have then shown that two blocks of the system compensate each other, leading to a lower sensitivity to TID. From this analysis, testing at system level is discussed.

2. Experiment Description

2.1. System under Test (SUT)

The system under test is a PoL converter developed by 3D-Plus. It is a custom-built system based on COTS components, with 11 references of active components, (6 discrete and 5 integrated circuits) from different technologies (bipolar/CMOS). PoL is a space qualified product available as 3D system-in-package (SiP) module, characterized (based on component-level tests [14]) up to a TID level of 50 krad. In our work, the original 3D SiPs as well as 2D prototype boards were tested. The 2D board has size of 85 mm × 95 mm and is functionally and electrically equivalent to the 3D SiP. Components are on a single layer for the 2D board whereas components of the SiP are distributed on 3 layers, one above another, and encapsulated in a 26.5 mm × 25 mm × 10 mm metalized package. Another difference is that two CMOS ICs on the 2D board have different date codes than for the 3D SiP (but have the same reference and come from the same manufacturer). The PoL functional block diagram is presented in Figure 1. The photograph of the PoL 3D SiP and 2D boards is given in Figure 2.

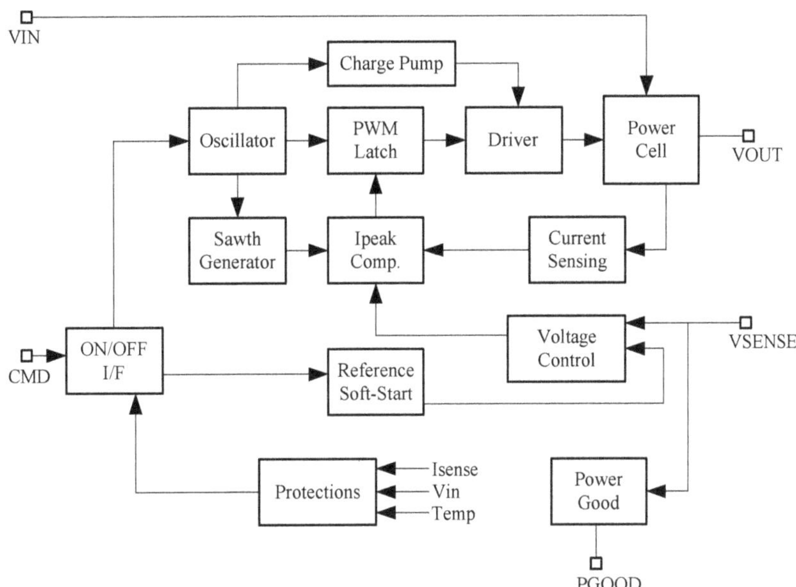

Figure 1. PoL functional block diagram.

2.2. Test at Component Level

TID tests on the 11 references of basic devices were characterized [14] during test campaigns performed in accordance with ESA Standard ESCC-22900.

Irradiation was performed using the Co^{60} source GIF at the Université Catholic de Louvain (UCL) facility in Belgium. The dose rate was between 100 and 360 rad(Si)/h. The bias conditions during exposure are the worst-case bias as per the PoL design justification document.

It was shown in [14] that all the irradiated devices are functional up to 50 krad. In a first step, parameter drifts under irradiation were evaluated regarding the tolerances given by the manufacturer datasheet. In a second step, if drift was higher than the initial tolerance, a specific analysis was performed to evaluate the acceptability of the variation in the PoL system. A summary of the total dose test results is given in Table 1.

Figure 2. PoL 3D SiP on the auxiliary board and attached to the motherboard (first plan) and two PoL 2D boards on their motherboards (second plan).

Table 1. TID status for active devices used in PoL module (from [14]).

Type	Results in Accordance with Manufacturer Datasheet	Results in Accordance with PoL Design Requirement
Diodes	>50 Krad	>50 Krad
Transistors	>50 Krad	>50 Krad
Voltage reference	15.4 Krad	>50 Krad
Logic gates	7 Krad	>50 Krad
	5 Krad	>50 Krad
Comparator	>50 Krad	>50 Krad
Op-amplifiers	9.8 Krad	>50 Krad
	4.9 Krad	>50 Krad

From the TID test campaign and results analysis according to the acceptable drifts for the PoL system, the PoL converter has been guaranteed for a total dose of 50 krad(Si).

2.3. Test at System Level

All SUTs were supplied during irradiation (5 V); the output voltage of each PoL (VOUT on Figure 1) was set to 2.5 V and the load connected to VOUT was 8 Ω. After each irradiation step, each SUT was removed from the irradiator, reconnected to the power supply and characterized electrically: 29 parameters were measured for 2D boards and 33 for 3D SiP, and both DC and AC parameters were measured. In this paper, only the most relevant experimental results are presented.

2.3.1. Irradiation with a Co^{60} Source

In order to achieve a direct comparison with the tests performed at component level in [14], complete PoL systems were irradiated using a Co^{60} source. One PoL 3D SiP

(SUT022) and one PoL 2D (SUT012) board were irradiated using the Co^{60} source of the PRESERVE platform at the University of Montpellier. Irradiation was performed up to 125Krad(Si) and the dose rate was 15 rad/h.

2.3.2. Irradiation with an X-ray Source

One PoL 3D SiP (SUT020) and two PoL 2D boards (SUT010 and SUT011) were irradiated using the X-RAD 320 irradiator of the PRESERVE platform at the University of Montpellier, with potential of the X-ray tube as high as 320 kV/12.5 mA. The dose rate used was 50 krad/h for all 3 SUTs tested.

3. Experimental Results and Analysis

3.1. Test at System Level Using a Co^{60} Source

At 118 krad(Si) both SUTs were fully functional. Among the 29 parameters measured, few of them drifted and the variation remained very small and largely inside the specifications. In Table 2, the parameters presenting the largest variations at 118 Krad(Si) are reported with their percentage of variation compared to the prerad value: the ON supply current, the OFF supply current and the oscillator frequency.

Table 2. Percentage of variation compared to the prerad value of parameters at 118 Krad(Si) (Co^{60} irradiation).

	Percentage of Variation Compared to the Prerad Value	
	3DSip Module	**2D Module**
ON supply current	0%	1.2%
OFF supply current	6%	25%
Oscillator frequency	1.5%	2.8%

The OFF supply current was measured when all the integrated circuits were supplied but the oscillator was stopped, therefore the loop controlling the output power was stopped and VOUT = 0V. The ON supply current was measured when the PoL system supplied the output.

The important point is then that the PoL system, when evaluated at system level, was shown to reach 118 krad(Si), whereas testing at component level allowed its qualification only at 50 krad(Si).

It is also important to note that the dose rate used for the test at system level was at least 7 times lower than the one used for the test at component level. The 15 rad/h used is compliant with testing for systems sensitive to ELDRS [13]. This also shows that no significant ELDRS is at play in the global system functionality.

The higher total dose reached during the test at system level than that using the test at component level could imply that there is compensation, at circuit level, between some of the system's blocks (Figure 1).

3.2. Test at System Level Using X-rays

In order to evaluate the limit of the system and to obtain significant variations of parameters to attempt an analysis of the possible interactions between the different system blocks, irradiations were performed using X-rays at a 50 krad/h dose rate. The goal was not to compare X-rays and Co^{60} irradiations, but to reach, in a reasonable time, a total ionizing dose high enough to observe the system's degradation.

The output voltage of the PoL (VOUT) is reported as a function of TID in Figure 3. For both 2D boards (SUT010 and SUT011) functional failures were observed, respectively, at 450 and 500 krad (at the end of the irradiation step, the system was disconnected and then connected to the test bench but did not power up). This means that the last doses for which the device was functional were, respectively, 400 and 475 krad for SUT 010 and SUT 011. It is important to note that for the 3D SiP device, no functional failure was observed (the irradiation was stopped at 500 krad). This is assumed to be related to the SiP packaging that mitigates X-rays on the internal circuit. From Figure 3, we can observe a very small drift of VOUT. The maximum measured drift of VOUT (which is the crucial parameter for the user) was not different from the initial value by more than 0.4% (SUT020, 500 krad) and the drift observed for SUT010 and SUT011 just one irradiation step before they failed was less than 0.3%. In Figure 3, two behaviors might be observed: an increase of the measured values up to level of 50–100 krad and a decrease above 50–100 krad. However, the goal of presenting the curves in Figure 3 was to show that the observed change of the output voltage was insignificant during irradiation and could not be used to predict that the system would soon fail. In this context, the insignificant linear change of the VOUT observed for TID above 50 krad gives no advantage in predicting when the system may fail.

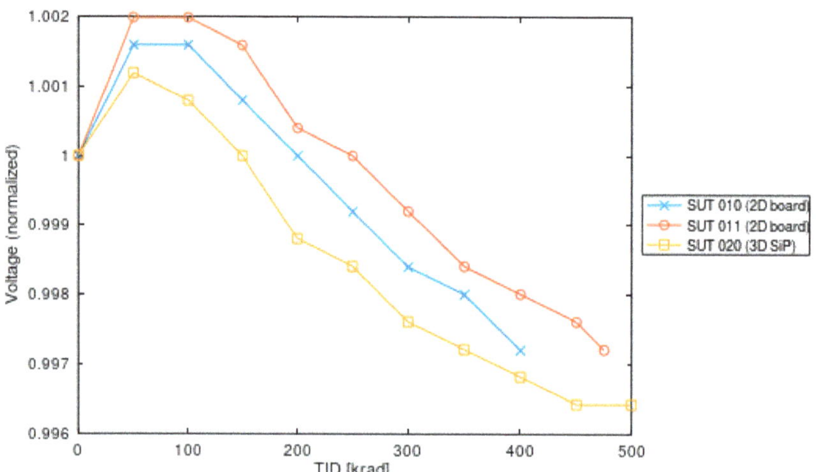

Figure 3. Output voltage (normalized) of SUTs as a function of the total ionizing dose (X-ray irradiation). SUT 010 and SUT 011 failed, respectively, at 450 and 500 krad (measurements performed at these two TID steps have shown that output voltage is ~0 V and that most of the internal subsystems are not working). The 3D SiP was still fully functional.

In Table 3, the percentages of variation compared to the prerad values are given at 400 krad(Si) for the ON supply current, the OFF supply current and the oscillator frequency to be compared to values shown in Table 2. No other parameters have shown a significant increase.

Table 3. Percentages of variation compared to the prerad values of parameters at 400 krad(Si) (X-rays irradiation).

	Percentage of Variation Compared to the Prerad Value	
	3DSip Module	**2D Module**
ON supply current	1%	12.5%
OFF supply current	59%	100%
Oscillator frequency	12.5%	18%

It is shown that the OFF supply current strongly increases for both 2D and 3D systems, by 100% and 59%, respectively. The ON supply current increase is important for the 2D board (12.5%).

In Figure 4, the normalized oscillator frequency value is represented as a function of TID for 2D boards and 3D SiP systems. For the 2D boards (SUT010 and SUT011), the oscillator frequency value increases by a factor of 1.18. A saturation of the degradation is observed after 150 krad. The saturation value is equal to 490 kHz. The 3D SiP system also shows an increase in the oscillator frequency value by a factor of 1.125. A saturation is observed after 250 krad. The saturation value is equal to 440 kHz. Notches observed in each curve (e.g., at 200 krad for SUT 010 curve) are due to annealing, when irradiation had to be stopped for longer period (e.g., whole night).

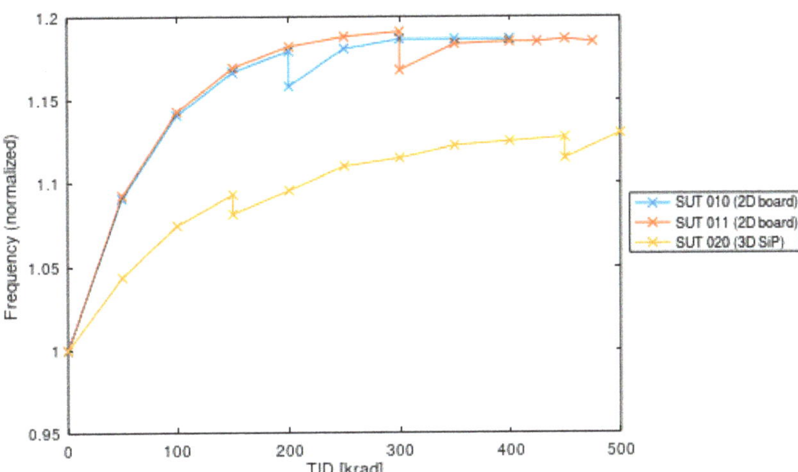

Figure 4. Normalized oscillator frequency value as a function of TID for 2D boards and 3D SiP. (notches observed in each curve (e.g., at 200 krad for SUT 010 curve) are due to annealing, when irradiation had to be stopped for a longer period (e.g., whole night)).

Considering the results for 2D boards, when the PoL system was designed, specifications were given for the irradiation tests at device level. The specification limit for the oscillator frequency was 450 kHz. Results presented in our work show that oscillator frequency is 8–9% higher than this limit as soon as the total dose reaches 150 krad but the PoL remains functional up to a total dose above 400 krad.

During component-level qualification of PoL [14] the Schmitt trigger, being a core part of the oscillator, was irradiated up to 50 krad with the Co^{60} source. Acceptability of Schmitt trigger variations has been verified from results at component level using standards but also on the PoL system. The maximum frequency drift measured for the oscillator built with an irradiated component was from 419 to 427 kHz and the part was then accepted. If, during this component-level qualification, the degradation of frequency had been similar to the X-ray irradiation test shown in Figure 4, the part would have been rejected, but the system-level test (Figure 3) shows that, regardless of the degradation of the oscillator frequency, the system remains functional not only after 50 krad, but also up to 400 krad under X-rays and 118 krad under Co^{60} irradiations.

3.3. Tracking of the System Failure Source

From all the measurements made, there is no obvious way to identify the process at play leading to the 2D board failures above 400 krad. Investigation of the failure source of 2D boards tested with X-rays was performed and the conducted analysis is presented hereafter.

Increased current consumption in ON mode (Table 3) suggested that switching time of the power MOSFET might be longer—this was confirmed by analyzing waveforms of the MOSFET driver signal recorded at each step of the irradiation (see Figure 5 for example waveforms recorded for SUT010).

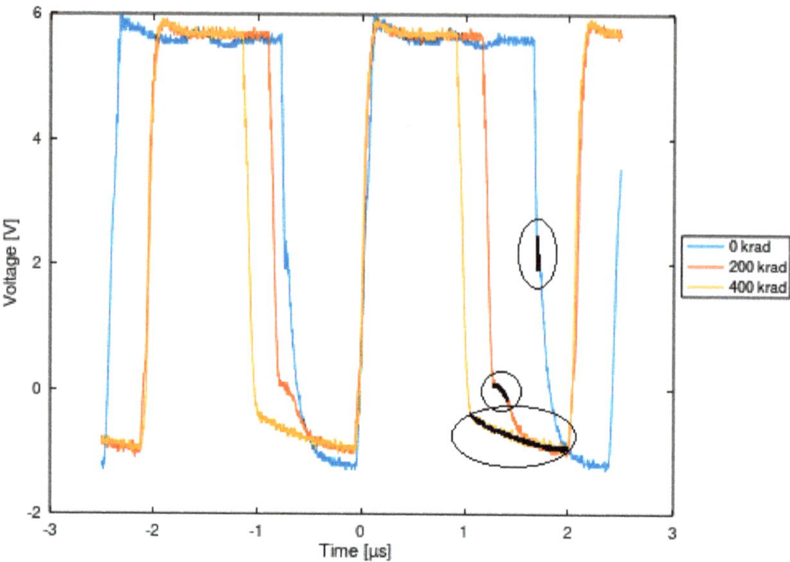

Figure 5. SUT010 MOSFET driver waveforms recorded at steps 0, 200 and 400 krad (X-ray irradiation). Black color marks switching time for each irradiation step. Switching time is approximately 3 times longer for 200 krad (than prerad) and more than 15 times longer for 400 krad (than prerad).

These waveforms show degradation of the switching threshold and increase of switching time of the MOSFET. For a high enough degradation, the MOSFET does not open fully (ON resistance is high), therefore it does not provide enough current to the PoL output. Because this current is too low, the feedback loop drives the control voltage too high and the overcurrent protection (OCP, based on the voltage control loop signal) is triggered. This triggering of the OCP is observed during characterization of failed SUTs (SUT010 at 450 krad and SUT011 at 500 krad) (Figure 6). The OCP signal not only provides information about the overcurrent state, but is also the input to the auto-reset circuit of the PoL. During normal operation of the PoL, when there is no overcurrent state, this signal is at the level of ~0.05 V. We can see from Figure 6 that the value of the OCP signal is repeated in the cycle: (1) slow decrease from ~3 V to around 0.5 V—in the beginning the OCP is in a high logic state, forcing the reset of the PoL; (2) at some point, the OCP stops resetting the PoL and the converter starts to operate; (3) the operation of the PoL leads again to the overcurrent state—the OCP signal from the level of ~0.5 V goes rapidly to the level of ~3 V. The PoL is again reset.

The switching time variations as a function of the total ionizing dose are shown in Figure 7 for both 2D boards and 3D SiP systems. It is shown that both 2D boards show a significant increase of the switching time compared to the 3D SiP systems. Such a difference is assumed to be due to the SiP packaging that reduces the total ionizing dose received by the components under X-ray irradiation. It is also shown that both 2D boards have the same degradation of the switching time up to 250 krad, but for doses above 250 krad, the SUT 010 systems present a higher degradation of the switching time than the SUT020 system for a given dose.

Considering the dose leading to the system failure (Figure 3), for SUT010 failure is observed after 400 krad and for SUT011 failure is observed after 475 krad. In both cases, failure is observed when a 1 μs value of the switching time is reached. We then have a direct correlation between the failure of the system and the switching time parameter that allows us to identify the process at play leading to the device failure.

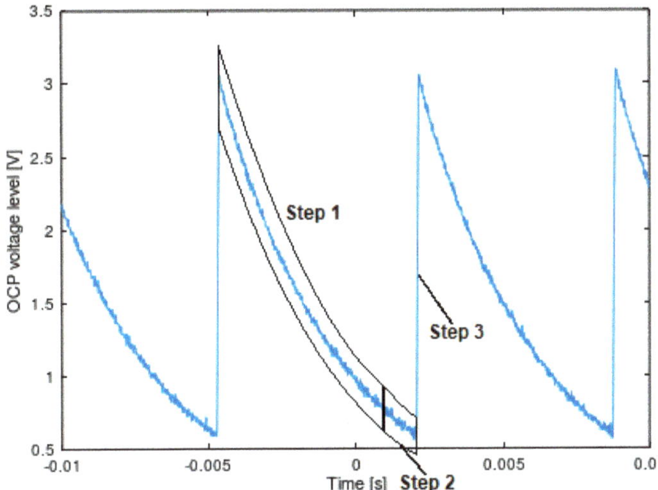

Figure 6. OCP signal as observed after the failure of the SUT010. 3 different phases/steps of operation of the PoL during failure are marked: (1) PoL is in the reset state and the OCP signal decreases; (2) OCP signal enters the range of logic low state and the PoL starts to operate; (3) OCP signal is triggered and the PoL is reset again.

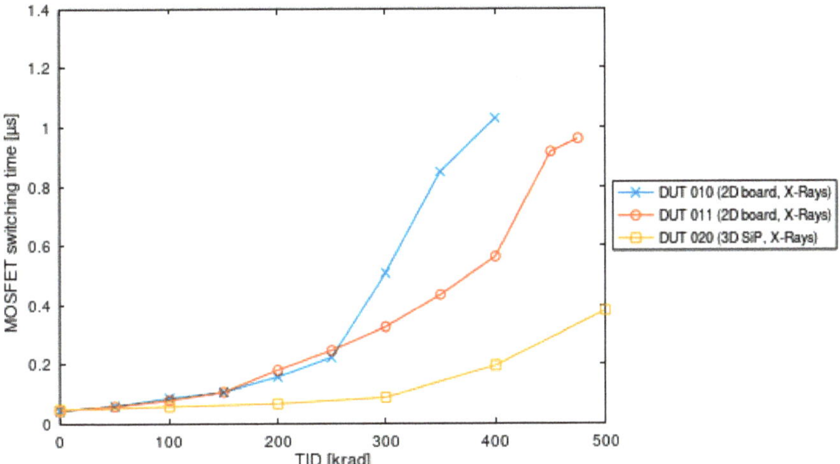

Figure 7. Switching time as a function of the total ionizing dose for both 2D boards and 3D SiP systems.

Let us now consider the fact that the system survived to a total dose significantly higher than 50 krad, which was the qualified level after the test at component level. Although the system failure was finally tracked down to the failure of a single component, it is worth noting that several parameters of the system were changed because of the MOSFET degradation: current consumption, parameters of the voltage control loop (compensating MOSFET degradation to retain a good level of VOUT) and MOSFET driver fall time. Particularly, it might be observed that the duty cycle of the MOSFET driver is changed by the PoL feedback mechanism, in order to compensate for the degradation of the MOSFET (see Figure 8). The related effects are presented in the block diagram in Figure 9: (1) the reduced output voltage is detected by the voltage control block; (2) the correction signal is sent to the PWM generation block; (3) the duty cycle of the MOSFET driver signal is modified.

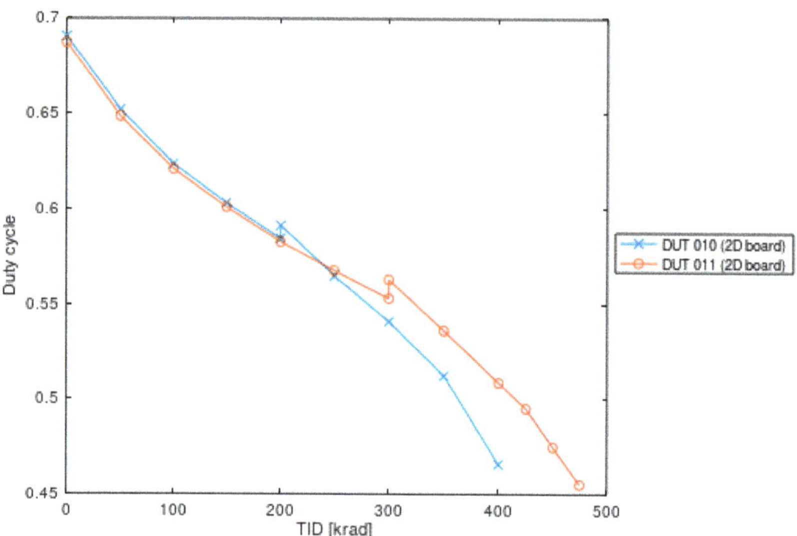

Figure 8. PoL 2D board MOSFET driver duty cycle change as a function of the total ionizing dose (X-ray irradiation).

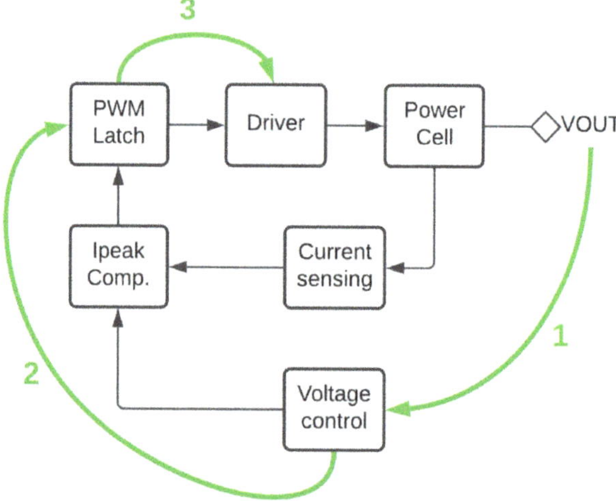

Figure 9. Block diagram explaining compensation of the MOSFET degradation. 3 effects are marked on the diagram: (1) the reduced output voltage is detected by the voltage control block; (2) the correction signal is sent to the PWM generation block; (3) the duty cycle of the MOSFET driver signal is modified.

In other words, the MOSFET degradation is compensated by the voltage control loop, extending the TID tolerance of the system.

4. Discussion

Testing at system level allows us, for this specific system, to show that the system can reach a higher total ionizing dose before failure than a test at component level, but the testing at system level could also be more difficult to qualify a system.

In our case, the system was shown to be more tolerant to TID, but if a failure had been observed during this test before 50 krad, the difficulty would have been to find the origin of the failure in order to improve the system and make corrections to the design. It is then very important to have a high observability during the testing, which means there are many system parameters to monitor. In our case, such a strong observability was possible due to the fact that SUT was a "white box". Full design information, as well as the placement of specific test points, were known during test preparation and during SUT characterization, which may not always be the case for other systems (e.g., COTS systems).

It was shown (Figure 7) that the two 2D boards have different kinetics for switching time degradation. SUT010 degrades more rapidly than SUT011. Such a difference is assumed to be induced by the part-to-part variations of the basic devices. Once again, in our case, this is not a problem as the system is shown to be tolerant for high total doses, but this will not be always the case. If we consider that the PoL system should be qualified for a 450 krad total dose, the difference between the two 2D boards generates an uncertainty as SUT010 will pass the test whereas SUT011 will not. A statistical approach would then be necessary in order to minimize the error induced by the component part-to-part difference.

It is also important to consider if the test performed at system level is conservative or not. In our case, the failure was shown to have been induced by the increase in the switching time leading to a decrease in the output current. This will then depend on the load at the output of the PoL. It is then expected that by decreasing the load (8 ohms during our test), the degradation will also increase, leading to a lower tolerance to TID. Our test was not performed in the worst-case conditions, but defining the worst-case conditions was difficult when the failure source at system level was not known before the test. Such a remark goes in the direction towards testing at system level "as your mission". The obtained results might be also interpreted in the way that for the given system configuration (i.e., output voltage, output load), there was a margin given by the design that could be used to increase tolerance to TID. For another configuration (particularly the worst-case configuration), there might be not as large a margin to be used.

5. Conclusions

One of the advantages of the testing at system-level approach is the possibility to obtain more realistic results because the influence of the component's degradation on the system health is already included in system-level test results. In order to evaluate such an approach, experiments were performed on a system-in-package point-of-load converter to evaluate its sensitivity to the TID. Two different approaches were used. The PoL converter was evaluated through tests at component level using standards and by a complete system-level approach.

The main results show that using standards leads to a qualification of the PoL at 50 krad(Si), whereas testing at system level shows that the PoL is fully functional at 118 krad(Si). Both approaches were performed using a classical Co^{60} source.

The test at system level was stopped at 118 Krad(Si) for experimental duration reasons and the PoL was still fully functional. X-rays were then used to go further in a short time irradiation. The TID level observed was higher than 400 krad before system failures.

Analysis of the results has shown that the extension of the TID tolerance of the system is related to compensations between some electrical blocks in the complete system.

In the discussion, it was clearly shown that the results obtained for this particular system cannot be generalized to others. The results were shown to represent an ideal case when testing at system level and points that have to be considered in a more general case were raised. The first point is related to the observability when performing tests at system level; the second one deals with the part-to-part variations; and the last one relates to the difficulty in defining the worst-case conditions when irradiating at system level.

Author Contributions: Conceptualization, T.R., F.S. and P.-X.W.; methodology, T.R., F.S. and P.-X.W.; software, T.R.; validation, T.R., F.S. and P.-X.W.; formal analysis, T.R. and F.S.; investigation, T.R., F.S., P.D. and P.-X.W.; resources, T.R., F.S., K.N., J.B., T.M., P.K. and P.-X.W.; data curation, T.R.; writing—original draft preparation, T.R. and F.S.; writing—review and editing, T.R., F.S., K.N., J.B., T.M., P.K., P.D., A.T., P.-X.W.; visualization, T.R.; supervision, F.S. and P.-X.W.; project administration, F.S. and P.-X.W.; funding acquisition, F.S. and P.-X.W. All authors have read and agreed to the published version of the manuscript.

Funding: X-rays and Co^{60} tests were performed using the PRESERVE Platform, which was funded by the Occitanie Region and the EU via ERDF funds. This work received funding from the European Union's Horizon 2020 research and innovation program under the Marie Skłodowska-Curie grant agreement number 721624.

Conflicts of Interest: The authors declare no conflict of interest.

References

1. Shaneyfelt, M.R.; Schwank, J.R.; Dodd, P.E.; Felix, J.A. Total Ionizing Dose and Single Event Effects Hardness Assurance Qualification Issues for Microelectronics. *IEEE Trans. Nucl. Sci.* **2008**, *55*, 1926–1946. [CrossRef]
2. Schwank, J.R.; Shaneyfelt, M.R.; Dodd, P.E. Radiation Hardness Assurance Testing of Microelectronic Devices and Integrated Circuits: Radiation Environments, Physical Mechanisms, and Foundations for Hardness Assurance. *IEEE Trans. Nucl. Sci.* **2013**, *60*, 2074–2100. [CrossRef]
3. Schwank, J.R.; Shaneyfelt, M.R.; Dodd, P.E. Radiation Hardness Assurance Testing of Microelectronic Devices and Integrated Circuits: Test Guideline for Proton and Heavy Ion Single-Event Effects. *IEEE Trans. Nucl. Sci.* **2013**, *60*, 2101–2118. [CrossRef]
4. Rajkowski, T.; Saigné, F.; Pouget, V.; Wrobel, F.; Touboul, A.; Boch, J.; Kohler, P.; Dubus, P.; Wang, P.X. Analysis of SET Propagation in a System in Package Point of Load Converter. *IEEE Trans. Nucl. Sci.* **2020**, *67*, 1494–1502. [CrossRef]
5. De Bibikoff, A.; Lamberbourg, P. Method for System-level testing of COTS electronic board under High Energy Heavy Ions. *IEEE Trans. Nucl. Sci.* **2020**, *67*, 2179–2187. [CrossRef]
6. Fernández, G.; Bárcena, J.; Muñoz, E. TID test on ITER Interlock Discharge Loop Interface Box (DLIB) system, an example of radiation test at equipment level. In Proceedings of the 17th European Conference on Radiation and Its Effects on Components and Systems (RADECS), Geneva, Switzerland, 2–6 October 2017.
7. Rousselet, M.; Adell, P.C.; Sheldon, D.J.; Boch, J.; Schone, H.; Saigné, F. Use and benefits of COTS board level testing for radiation hardness assurance. In Proceedings of the 2016 European Conference on Radiation and Its Effects on Components and Systems, Bremen, Germany, 19–23 September 2016.
8. Virmontois, C.; Belloir, J.; Beaumel, M.; Vriet, A.; Perrot, N.; Sellier, C.; Bezine, J.; Gambart, D.; Blain, D.; Garcia-Sanchez, E.; et al. Dose and Single-Event Effects on a Color CMOS Camera for Space Exploration. *IEEE Trans. Nucl. Sci.* **2019**, *66*, 104–110. [CrossRef]
9. Secondo, R.; Alia, R.G.; Peronnard, P.; Brugger, M.; Masi, A. System level radiation characterization of a 1U CubeSat based on CERN radiation monitoring technology. *IEEE Trans. Nucl. Sci.* **2018**, *65*, 1694–1699. [CrossRef]
10. Coronetti, A.; Manni, F.; Mekki, J.; Dangla, D.; Virmontois, C.; Kerboub, N.; Garcia Alia, R. Mixed-field radiation qualification of a COTS space on-board computer along with its CMOS camera payload. In Proceedings of the 2019 European Conference on Radiation and Its Effects on Components and Systems, Montpellier, France, 16–20 September 2019.
11. Coronetti, A.; Alia, R.G.; Budroweit, J.; Rajkowski, T.; Lopes, I.D.C.; Niskanen, K.; Soderstrom, D.; Cazzaniga, C.; Ferraro, R.; Danzeca, S.; et al. Radiation hardness assurance through system-level testing: Risk acceptance, facility requirements, test methodology and data exploitation. *IEEE Trans. Nucl. Sci.* **2021**. [CrossRef]
12. Wei, H.; Yueke, W.; Kefei, X.; Wei, D.; Pengcheng, L. Total ionizing dose vulnerability analysis and on-orbit prediction for spaceborne signal processing platform. In Proceedings of the IEEE International Conference on Signal and Image Processing (ICSIP), Beijing, China, 13–15 August 2016.
13. Total dose steady-state irradiation test method. In *ESCC Basic Specification*; No. 22900; European Space Agency: Paris, France, 2016; Issue 5.
14. Vassal, M.; Dubus, P.; Fiant, N. Electrical performances and radiation qualification tests results of a highly integrated and Space qualified Point of Load Converter. In Proceedings of the 2011 European Conference on Radiation and Its Effects on Components and Systems, Seville, Spain, 19–23 September 2011.

electronics

Article

Radiation Qualification by Means of the System-Level Testing: Opportunities and Limitations

Tomasz Rajkowski [1,2,*] , Frédéric Saigné [2] and Pierre-Xiao Wang [1]

1   3D PLUS, 408 Rue Hélène Boucher, 78532 Buc, France; pwang@3d-plus.com
2   IES, University of Montpellier, CNRS, 34000 Montpellier, France; frederic.saigne@ies.univ-montp2.fr
*   Correspondence: trajkowski.openaccess@gmail.com

Abstract: System-level radiation testing of electronics is evaluated, based on test examples of the System-in-Package (SiP) module irradiations. Total ionizing dose and single event effects tests are analyzed to better understand the opportunities and limitations of the system-level approach in the context of the radiation qualification of electronics. Impact on the SiP product development is discussed.

Keywords: system-level tests; radiation hardness assurance; total ionizing dose; single event effects

Citation: Rajkowski, T.; Saigné, F.; Wang, P.-X. Radiation Qualification by Means of the System-Level Testing: Opportunities and Limitations. *Electronics* **2022**, *11*, 378. https://doi.org/10.3390/electronics11030378

Academic Editor: Raffaele Giordano

Received: 15 November 2021
Accepted: 25 January 2022
Published: 27 January 2022

Publisher's Note: MDPI stays neutral with regard to jurisdictional claims in published maps and institutional affiliations.

1. Introduction

In recent years, we could observe a huge increase in the number of small/micro-/nano-satellites that were launched and that are built of non-space grade electronics, but from COTS (commercial off the shelf) components. COTS components are manufactured without following as strict procedures (e.g., testing and inspection) as space-grade electronic components. In addition, the traceability of COTS components is very often limited, giving the end-user reduced knowledge on the manufacturer and manufacturing process of the product. The use of COTS is primarily driven by the requirement (and possibility) to reduce the cost and development time of missions. In addition, COTS components may enable much higher performance than available with the use of space-grade electronics. However, to assure mission success, radiation hardness assurance (RHA) schemes need to be applied [1], otherwise, the risk induced by the space radiation effects on the COTS-based systems might be out of control. This risk is related to the radiation environment in space, i.e., particles trapped by the Earth's magnetic field, Galactic Cosmic Rays and Solar Particle Events. The space radiation may be a source of cumulative effects (such as total ionizing dose (TID) or displacement damage) and single event effects (SEE), and in general may lead to decreased performance of specific functions, temporal malfunctions or critical failure and loss of mission. In this context, the goal of RHA is to ensure that all the potentially radiation-sensitive units of a space system (including the space system itself) will meet their design specifications up to the end of the targeted mission [1]. The core part of the RHA is defining the target environment, defining the requirements and evaluating the design and components for the mission [2]. The standard RHA approach requires component-level testing of all the potentially sensitive components, which is time- and money-consuming. Therefore, simplified approaches are searched for, including testing of the reduced number of components in the system (e.g., the most critical components only) and system-level testing [3].

In this paper, the system-level radiation tests are discussed from the point of view of potential electronic system radiation qualification. The main goal of this paper is to present opportunities and limitations related to the radiation qualification of the System-in-Package (SiP) modules (and systems in general) using system-level tests.

While discussing the system-level radiation testing of electronics, it is essential to define what is understood as an electronic system and what is not. It was proposed in [4] to define a component as any device that cannot be physically partitioned without damaging it; therefore, anything manufactured on a chip falls into this category. Furthermore, a "system" would be something at a higher integration level than a component (PCB (printed circuit board), assembly of PCB boards, SiP module, satellite subsystem or even whole satellite). Definitions proposed in [4] were found helpful in the context of radiation testing and RHA and will also be used within this description and this project.

SiP is often used as a commercial name for a product, the electronic system developed in 3D PLUS on-demand of a client, based on provided part list, block diagram, required functionality/features list, etc., enclosed in a single, miniaturized package (typically not bigger than $3 \times 3 \times 2$ cm^3). In this work, SiP will refer to a more general meaning of the electronic system consisting of a number of dissimilar integrated circuits and passive components, enclosed in a single, miniaturized package. However, the SiP mostly discussed in this paper is the Point-of-Load converter (PoL) from 3D PLUS [5], as all the radiation test experiments performed within this study had the PoL as the test vehicle.

The system-level test might be defined as the one in which parameters and/or functionality of the whole system is monitored at the time of the test, instead of parameters of the specific component only. However, not the whole system has to be irradiated at a given time, to call it a system-level test. For example, during laser test, the laser is focused only on one integrated circuit at a time, but if the component is a part of the working system and parameters of the system are monitored during the test, it is a system-level test [6]. Similarly, irradiation with heavy ions of a single component in the monitored system (such as described, e.g., in [7]) is also a system-level test.

Different approaches for system-level testing were already used and discussed in the literature. A summary of these approaches is presented in Table 1:

Table 1. Summary of the system-level test approaches discussed in the literature.

Test Type	Radiation Source/Test Approach	Main Features
TID	Photons—gamma rays (Co-60 ($\approx$1.2–1.3 MeV), Cs-137 ($\approx$0.7 MeV)) and X-ray [8–10]	- high penetration of the source (particularly Co-60) - high capabilities in testing big systems (limited for X-ray due to typical geometry of X-ray machines)
SEE	protons [11–13]	- high penetration of particles - broad beams available, suitable for board/system-level testing - well-established test method for 200 MeV protons, with provided formulas to calculate SEE rate for the orbit of International Space Station - limited linear energy transfer (LET) (LET of secondary particles after proton hit at the component die)
	heavy ions: standard (<10 MeV/n) and high (10–100 MeV/n) energy	- limited penetration of particles (opening of the component packages required: only board-level testing possible, not for bigger systems) - test in vacuum for standard energy heavy ions - high LET, up to 100 MeV·cm^2/mg
	heavy ions: very high energy (0.1–5 GeV/n) [14]	- moderate penetration, suitable for board-level testing, not for bigger systems - high LET; however, for board-level testing with homogenous LET limited to 30 MeV·cm^2/mg for two-side PCB testing, slightly more than 40 MeV·cm^2/mg for one-side PCB testing,
	heavy ions: ultrahigh energy (5–150 GeV/n) [15,16]	- high penetration, possible to test whole systems or even several systems stacked in the beam line - limited LET: around 10–15 MeV·cm^2/mg
mixed	CHARM facility: high energy hadrons (HEH, with energies up to 24 GeV) [4,17–20]	- an alternative approach for RHA: to test with HEH spectrum equivalent to the one at the target environment - high penetration of the field, big space for testing of systems - a useful tool for soft error rate estimation, however, LET is limited to around 15–17 MeV·cm^2/mg

The table gives references to documents where different system-level test approaches are discussed. For more detailed information on performing tests in different facilities, or using different approaches, the Reader is encouraged to consult the references provided. Another relevant reference is the first guidelines document for system-level testing for space systems, established last year [21].

Some of the system-level test approaches presented in Table 1 were used to test the PoL. In this paper, selected results from these experiments will be presented, with the main goal to discuss opportunities and limitations related to the radiation qualification of the SiP modules using system-level tests.

2. Overview of System-Level Tests of the PoL SiP

The test vehicle chosen for this study was the PoL module from 3D PLUS and different entities of this SiP (both 2D prototype boards and 3D modules) were irradiated during experiments discussed in this paper. The PoL is a custom-built system based on COTS components, with 11 references of active components (6 discrete and 5 Integrated Circuits), from different technologies (bipolar/CMOS). More details on the PoL were already given in another paper in this journal issue [22], but the PoL schematic is reminded in Figure 1:

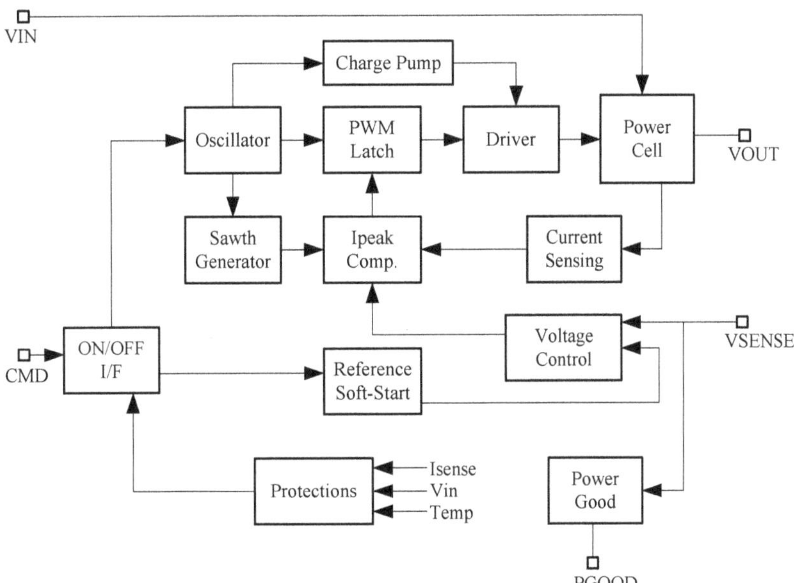

Figure 1. PoL functional block diagram.

PoL is a space qualified product, characterized with the use of component-level tests up to TID level of 50 krad and LET level of 80 MeV·cm^2/mg [7].

An overview of radiation tests of the PoL performed within this study is given in Table 2. TID and SEE tests are presented shortly, as well as the mixed-field tests at the CHARM facility.

Table 2. Overview of radiation tests of the PoL.

Test Type and Radiation Source	Main Results
SEE: high energy heavy ions (GANIL facility)	- 2 units of the PoL 2D boards were irradiated - sensitive chips of the system were irradiated with high LET ions (up to 94.3 MeV·cm^2/mg) and the system remained functional - SETs at certain nodes, also propagating to the system output were observed, with no impact on the main system functionsmore results in [23]
The mixed field at CHARM (CERN)	- 2 units of the PoL 2D boards were irradiated - test reproduced the environment of the example application: LEO orbit; no failures were observed - the main impact of the reproduced radiative environment on the system behaviour was the SETs at the voltage reference chip (with no impact on the system main functions) and slight degradation of the key system parameters (supply current: ~2% drift, output voltage: ~1% drift) due to cumulative effects - more results in [24]
SEE: ultrahigh energy heavy ions (CHARM, CERN)	- 2 units of the PoL 2D boards and one 3D module were irradiated - parallel SEE test of 3 SUTs stacked in the beam line; the LET is homogenous but low - thousands of SETs were captured and some rare events were observed: the propagation of SET with a specific signature through one of the critical integrated circuits, forcing the device to power cycle - more results in [6,25]
Laser tests (Univ. Montpellier, PRESERVE platform)	- 1 unit of the PoL 2D board was irradiated - test with the good observability and possibility to reproduce rare effects—it was used to investigate rare events observed at CHARM - vast statistical data on the signature of the SET forcing power cycling of the system let us understand the propagation and power cycling mechanism - more results in [6]
SEE: standard energy heavy ions (CYCLONE facility at Université Catholique de Louvain)	- 1 unit of the PoL 2D board was irradiated - further investigation on the rare events observed at CHARM - test was used to confirm previous results from laser tests and SPICE simulations - more results in [6]
TID: Co-60 (Univ. Montpellier, PRESERVE platform)	- 1 unit of the PoL 2D board and 1 3D module were irradiated - SUTs irradiated up to 118 krad and are still functional - a test performed in the environmental worst case conditions (low dose rate), nevertheless the results of the system irradiation were much more optimistic than expected based on the component-level tests and qualification (the PoL product was qualified up to the level of 50 krad) - more results in [22]
TID: X-rays (Univ. Montpellier, PRESERVE platform)	- 2 units of the PoL 2D board and 1 3D module were irradiated - functional failures of 2D boards were observed at >400 krad, which is a much higher TID level of failure than expected from the product qualification - this high observability test could be used to track the complex failure mode of the system and to define the most sensitive component in the systemmore results in [22]

The next chapters of this paper shall present aggregated, cross-sectional analysis of test results from the seven test campaigns summarized in Table 2, with a special focus on the opportunities and limitations of the system-level testing.

3. Opportunities Given by System-Level Testing

The primary outcome of the system-level testing of electronics is the characterization (electrical, functional) of the whole system by a single test, i.e., the system-level test may provide:

- for TID tests: information on the degradation of observed parameters as a function of TID, information on the TID level that the failure of the system is observed,
- for SEE tests: energy/LET threshold of specific system-level or component level effect, a cross section of these effects (as a function of energy or LET),

- for mixed-field tests at CHARM: a number of events and information on the degradation due to cumulative effects in the environment defined by the high energy hadrons spectrum.

With these data, the user acquires some level of confidence in the system's radiation performance in a cost-efficient way. The amount and quality of data collected during system-level test will highly depend on the observability of system parameters provided by the test setup. Information from the system-level test might be used for design validation, or as an input for design improvements. Results have different specificity for TID and SEE test campaigns, therefore these two types of tests will be discussed separately.

3.1. Opportunities Given by TID Tests

Firstly, the opportunities resulting from performing the TID system-level radiation tests will be presented. These observations come from the Co-60 and X-rays irradiations performed on the PoL modules.

3.1.1. Margins Given by the System Are Incorporated in Test Results

One of interests in performing system-level testing is to evaluate margins given by the system. It is expected that the system may withstand a higher TID level than it results from component-level testing performed in the worst case conditions for specific components. This prediction is based on the fact that: (1) typically not all of the components in the system are working in their worst-case conditions, and (2) system designers apply safety margins (derating).

TID irradiations of the PoL allowed to observe these margins: it was observed that although degradation of the oscillator frequency is beyond design limits and beyond level obtained during component-level qualification of the core part of the oscillator, the system remains functional. Furthermore, although the system was qualified up to 50 krad based on Co-60 testing following the classical standard procedures [7], it could withstand 400 krad at X-ray (high dose rate) and at least 118 krad at Co-60 (low dose rate)—although, the test was not performed in the worst case conditions. Analysis of the results have shown that the extension of the TID tolerance of the system is related to compensations between some electrical blocks in the complete system—a broad description of this example was given in [22].

In general, it might be expected that system TID performance will be higher (or at least not lower) than the one derived from component-level testing and system-level testing is an interesting tool to estimate this overall performance. On the other hand, there is also a risk of overestimating system TID performance, if system tests are not performed in the worst-case condition—and if this condition will be met in the real use situation.

3.1.2. New Insight into Details of the System's Radiation Performance

Results of the system-level radiation tests may be the source of valuable information about the system details, and different PoL test campaigns also gave such information.

In the X-rays test (already mentioned in Section 3.1.1) it was possible to reach in relatively short time (around 8–10 h of irradiation) the total dose level high enough that the functional failure of two PoL modules was observed: above 400 krad and above 475 krad irradiation steps. More than 30 different electrical parameters were measured during each irradiation step of the TID tests, giving good observability during this experiment. Thanks to this good observability, it was possible to track the most sensitive component in the system (power MOSFET) and its most critical parameter, the MOSFET switching time (see Figure 2 for the recording of MOSFET driver waveforms in one of the irradiated PoL 2D boards and Figure 3 for the degradation of the MOSFET switching time in all three PoL modules irradiated with X-rays).

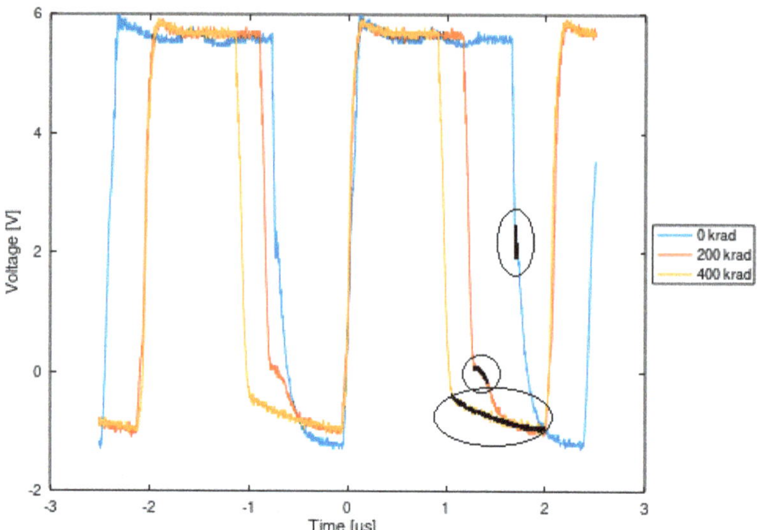

Figure 2. MOSFET driver waveforms recorded at steps 0, 200, 400 krad (X-rays irradiation of the PoL 2D board), after [22]. Black colour marks switching time for each irradiation step. Switching time is approximately 3 times longer for 200 krad (than prerad) and more than 15 times longer for 400 krad (than prerad). The device failed after the next irradiation step, i.e., it was not functional at the irradiation step of 450 krad.

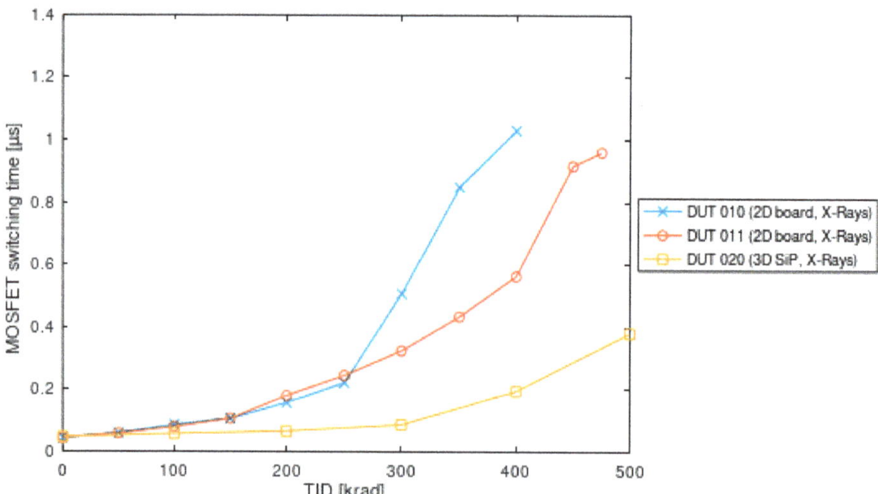

Figure 3. MOSFET switching time as a function of the total ionizing dose for two PoL 2D boards and one PoL 3D SiP module (after [22]).

This test has shown that system-level TID irradiation may be useful to (1) assess which parts should be replaced in the system to extend its radiation performance, (2) evaluate general radiation performance (TID level that the system may survive), (3) to define error signatures for the system. Such TID system-level test might be an interesting tool for verification of mature designs, but also evaluation at the early development phase of product/project.

3.2. Opportunities Given by SEE Tests

In this section, the opportunities resulting from performing the SEE system-level radiation tests will be presented. These observations come from the heavy ions, laser and mixed-field irradiations performed on the PoL modules (see Table 2 in Section 2).

3.2.1. Margins Given by the System Are Incorporated in Test Results

Some kind of margins might be also expected in the SEE system-level testing: masking of soft SEEs (such as single event upsets (SEUs) or single event transients (SETs)) or mitigation of some hard SEEs (such as single event latchup (SEL)) by protection mechanisms applied at system-level. Therefore, system-level SEE testing might also give more realistic results than tests at the component level because these masking and mitigation effects will be included in test results.

As an example of soft SEE masking, several test campaigns (e.g., heavy ion tests at GANIL, CHARM mixed-field, CHARM heavy ions, heavy ion tests at UCL, laser tests using the IES PRESERVE platform of the University of Montpellier) enabled to observe SETs at the voltage reference chip output, but most of these transients were not propagating to the general output of the device. These SETs were filtered out by the internal filter in the PoL circuit. An example of such a non-propagating transient (one of the several thousand similar events recorded during the CHARM UHE experiment) is presented in Figure 4 (left curve).

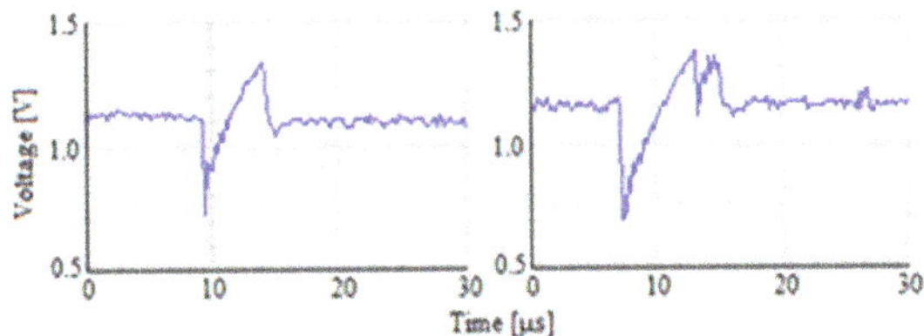

Figure 4. Example of SETs measured at voltage reference chip output, which caused power cycling (**left**) and which didn't cause power cycling (**right**) during PoL heavy ion test at CHARM facility.

Such results might be used for verification of the mitigation techniques efficiency in the system.

3.2.2. New Insight into Details of the System's Radiation Performance

Apart from the SET masking, there were also rare events observed during the CHARM UHE experiments, when SETs with the special signature (as on the right curve of Figure 4) were forcing power cycling of the device. The calculated cross-section for this event is between 1.3×10^{-8} and 2.3×10^{-8} cm^2, whereas the cross-section for SETs observed at the output of voltage reference chip is in order of ~10^{-5} cm^2. Detailed investigation of this phenomenon (presented fully in [6]) has shown that this specific signature of SETs is most probably due to double hits of ions at sensitive area of the chip, one hit after another shortly in time, therefore it is not something that should be expected in real operation conditions. Furthermore, it could be only observed after the significant degradation of the system due to cumulative effects, which was then leading to an increase of the voltage drop on the supply cables and a decrease of the supply voltage (as measured directly at PoL input). However, observation of such rare events enables to identify new failure modes (system-specific) and therefore to define error signatures for the system.

The presented example of the "rare event" observation shows that the system level testing may also provide information on synergistic effects at the system level: observed propagation of the SET was enabled by system degradation due to cumulative effects. Such effect had not been observed earlier and was very difficult to predict by analysis or component-level testing only. As for today, there is no procedure to evaluate synergistic effects and test at the system level seems to be the only way do to it.

3.3. Opportunities Given by System-Level Testing—Summary

Summarizing this section, it was shown based on different system-level radiation tests, that such tests might be useful in these areas:

- to verify the mitigation techniques efficiency,
- to identify new failure modes,
- to define error signatures for the system,
- to assess which parts should be replaced in the system,
- to evaluate general TID performance,
- to provide information on synergistic effects at the system level.

4. Limiting Factors of Qualification

Although there are interesting opportunities given by the system-level testing, performed experiments let also point out the crucial limitations in terms of system qualification by system-level test only. The most relevant results are presented in this section. Specific remarks for SEE and TID test cases will be given.

4.1. The Worst Case TEST Conditions

For the component-level tests, worst case configuration is well defined and described in standards [26,27] and literature [28,29]—both in terms of component configuration (such as applied bias) and in terms of environmental conditions (such as temperature or dose rate). In the case of system tests, the definition/choice of the worst-case conditions is more complex and might be the primary issue when considering qualification by only system-level tests.

4.1.1. Example of the Worst Case Test Conditions in TID Tests

As it was presented in the previous section, during system-level TID tests of the PoL it was possible to track the most sensitive component in the system and the most critical parameter, the MOSFET switching time. Analysis performed in [22] has shown that for the high enough degradation, the MOSFET doesn't open fully (ON resistance is high), therefore it doesn't provide enough current to the PoL output. However, this gives some clue that the system worst case might be dependent on the PoL output current configuration, i.e., it might be expected that for the higher output current (lower load resistance) the PoL would fail at a lower TID level because the MOSFET would not be able to give this higher current. On the other hand, it is possible that for the lower output current configuration, the TID failure threshold would be even higher than it was observed during the X-rays experiment.

4.1.2. Example of the Worst Case Test Conditions in SEE Tests

Another example mentioned in Section 3.2.2 was the SET propagation (during heavy ions test at CHARM), that was leading to power cycling of the PoL device. As it was already explained, this power cycling was only observed when the supply voltage of the PoL was significantly reduced. Then, we may see the dependence of the specific system-level effect from the power supply conditions: reduced supply voltage turns out to be the worst case condition for the propagation of these specific transients. Furthermore, this propagation was only observed for the specific signature of SETs (see Figure 4), which was produced by two particles passing the sensitive region of the voltage reference chip closely in time. Therefore, a high flux of these particles was needed to produce such an effect, which leads to another conclusion, that not only specific supply configuration, but also specific environmental condition (high particle flux) was needed to produce discussed effect.

4.1.3. General Remarks Regarding the Worst Case Test Conditions

Presented results may lead to the more general conclusions, that each system might have its own, specific worst-case configuration (such as internal state (ON/OFF/idle), input/output voltage, output load, clock frequency, number of system-specific operations per time unit, etc.) that is mostly related to the system internal design and might be difficult to predict without thorough analysis (acquiring more complex for more complex systems). In addition, it might be expected that for some systems only testing in different configurations will show what is the worst-case configuration. For some systems it may be possible to simplify this problem by testing in real conditions only, instead of worst-case conditions, therefore limiting the number of required tests. However, in many cases, e.g., for systems with many internal modes or functions, real conditions cannot be covered by testing in one configuration only.

On the other hand, environmental worst-case conditions might be different for different parts in the system (depending on the component technology), therefore it might be required to irradiate the system in different temperatures or dose rates to meet the worst-case condition for all the components onboard.

For example, when considering temperature, it is confirmed that for TID testing of components, both higher and lower temperatures might be the worst case, depending on the component type. For the SEE testing, the worst case condition for SEL is the high temperature, whereas for Single Event Burnout it is the low temperature. SEU testing is typically performed at room temperature, however, some specific IC designs may show up to have SEU rates dependent on the temperature [28]. Therefore, it may turn out that for complex designs, including different types of components, SEE testing should be performed over the full temperature range for the system. For TID, it should be sufficient to irradiate the system at one temperature and characterize it in different temperatures. In the example of PoL tests at CHARM mixed-field (see [24]), it was expected to observe SEL in the PoL with modified components—one of the components was known to be sensitive to SEL at LET as low as 1.2 MeV·cm^2/mg and its anti-latchup protection was removed. However, the latchup characterization of the component was performed at 125 °C and the test at CHARM was at room temperature, therefore not the worst case for SEL—this could be the reason why SEL was not observed during mixed-field tests at CHARM, although the high energy hadron fluence was as high as 3.47×10^{11} cm^{-2} and the CHARM secondaries typical maximum LET is 15 MeV·cm^2/mg.

4.2. Number of SUT Samples to Test

According to the component-level tests standards (such as ESCC 22900 and ESCC 25100), the number of samples required for qualification tests is 11 samples and 3 samples for TID tests and SEE tests, respectively, if the samples are from the same lot. For TID, it is required to test 5 samples biased, 5 samples unbiased and have one unirradiated control sample [27]. These recommendations result from the fact that there is always some part-to-part variability of parameters, even for the parts from the same lot, therefore the goal of tests is to obtain a statistical result.

4.2.1. Considerations on Number of SUT Samples in TID Tests

In our TID tests, such part-to-part variability was also observed. Let us take an example of the MOSFET switching time degradation during X-rays tests (see Figure 3). In all PoL modules that were tested, the MOSFET transistors were from the same production lot. Let us focus on test results for PoL 2D boards: even if the test configuration was the same, there might be observed faster degradation of the SUT010 MOSFET (blue colour) when compared to the SUT011 MOSFET (red colour), This difference is supposed to be the main factor explaining why the system-level degradation, and—finally—functional failure of the SUT010 device is observed at lower TID level than for SUT011. This result shows that the part-to-part variability might be transformed into the SUT-to-SUT variability: TID testing of only one SUT doesn't give full information about the TID performance of given device series, even if the components in the series are from the corresponding production lots. Testing of a higher number of SUTs is required to give reliable information on the TID radiation performance of the system.

On the other hand, if some risk might be accepted, one of the ways to reduce the required number of SUT samples would be to irradiate this reduced SUT sample set to a much higher TID level than expected in the final application of the system. In [9] it was chosen to irradiate the system (equipment) to the level 300 times higher than expected in the target application, to accept it for use. Although such margin is typically difficult to achieve for space applications, this approach might be sometimes feasible. If it is possible to irradiate the system up to the catastrophic failure and if it is possible to track the source of failure (such as in an example of X-ray tests [22]), this information might be also useful for further analysis of the TID level that system may withstand.

4.2.2. Considerations on Number of SUT Samples in SEE Tests

Considering system-level SEE testing, a recommendation from ESCC 22500 to test 3 samples (i.e., system units/instantiations in this case) seem to be appropriate for systems where parts are taken from the same production lots. The effect of each SEE on the system should be considered as independent from other SEEs, therefore such a system-level test might be considered as a simultaneous test of many integrated circuits inside of the system, incorporating part-to-part variability. Therefore, there is no need to perform tests on a higher number of SUT units, even if there is a higher number of integrated circuits in the system.

4.3. Irradiation Source Issues

Typical issues with the irradiation source to be used, are related to the beam penetration and the size of the SUT that might be irradiated with the homogenous beam.

4.3.1. Irradiation Source Issues in TID Tests

For TID tests, the Co-60 standard radiation source has high penetration and usually test area size is also accurate for even big systems. If another source is used, such as X-rays, constraints from the point of view of both penetration and available test area should be taken into account.

4.3.2. Irradiation Source Issues in SEE Tests

Beam parameters such as beam penetration range and beam size may be problematic particularly in SEE tests. Due to limited penetration of, e.g., heavy ions, component packages often need to be opened before irradiation (it was the case for the PoL tests at GANIL and UCL facilities). Sometimes chip dies need to be even thinned to assure that the ions reach the sensitive volume (s). Additional problems may arise when it comes to soldering the fragile component with the opened package on the PCB board. For system-level tests, it happens that it is not possible to irradiate all the components with the same LET value in one run, because components are on different layers of PCB. Furthermore, for systems that are not intended to be opened for testing (such as whole satellites or "black box" systems), penetration of the beam becomes a critical limitation for irradiating with high LET ions. (Considerations on the penetration of heavy ions in board-level tests, as well as related test methodology have been discussed in [14]). Another limitation comes from the particle beam size: it will be a problem with beam uniformity when irradiating larger components or systems, but for system-level testing, it may turn out to be impossible to irradiate the whole system at once and different parts of the big system shall be irradiated one after another. (See Tables 3 and 4 for the information on beam parameters of the chosen heavy ion and proton facilities.) However, the CHARM facility operating in mixed-field mode allows for the beam uniformity typically bigger than in standard facilities used for SEE testing.

Table 3. Beam parameters of the chosen heavy ion facilities (after [30]).

Heavy Ion Facility	Beam Size	Max Energy (MeV/u)	LET Range (MeV·cm^2/mg)
NSRL, USA	60 × 60 cm^2 (typically 20 × 20)	1500	0.5–94
RADEF, Finland	5 × 5 cm^2 (3 × 3 in air)	16.3	7–70
TAMU, USA	4.5 cm diameter	40	5–95
KVI, Netherlands	3 × 3 cm^2	90	4.5–65
UCL, Belgium	2.5cm diameter	9.3	1–62.5
GANIL, France	Few cm diameter, sweeping beam	60	5.4–97.6

Table 4. Beam parameters of the chosen proton facilities (after [30]).

	Energy (MeV)	Beam Size (cm Diameter)	Max. Range in Si (mm, SRIM2013)	Flux (p/cm^2/s)
PSI, Switzerland	6–230	9	176	<2 × 10^9
KVI, Netherlands	10–184	12	120	<10^9
UCL, Belgium	10–62	8	18	<5 × 10^8
RADEF, Finland	0.4–55	10	14	<3 × 10^8

4.4. Observability of Parameters

Observability of system parameters is important for monitoring of events during irradiation and assessment of the system health during or after irradiation. It is also crucial for tracking of failure root cause in the system (as it could be seen in the example of the PoL failing under X-ray: the source of failure could be tracked down to a failure of a single component thanks to high observability of parameters, see [22]).

One of the limitations in the observability is intrinsic for system-level testing, as it comes from the system integration. We can imagine, e.g., OpAmp being part of the complex circuit—it is not possible (or really hard) to measure OpAmp's bias current and many other parameters, while it's embedded in the system. Component-level testing, giving access to all the part's interfaces, is not restricted in this way and enables characterization of a broader number of parameters. However, limitation in the observability resulting from the system integration is not always the drawback, because not all of the parameters of the components in the system are important for monitoring, and usually, the system designer enables monitoring of the most important parameters by providing dedicated interfaces.

As it was discussed in Section 4.1, the radiation response of an electronic component may depend on the configuration used, such as supply voltage, operating frequency, etc. When testing at the component level, we are free to use any configuration that we want, because we are irradiating only a single part. When testing at the system level, we are limited to the configurations imposed by the system. It may be only one configuration or range of options to choose from but still limited by the design of the system. From this point of view, observability is reduced again by the system that not only limits access to component interfaces but also imposes specific allowed configurations that components may work—but this type of observability reduction should not limit capabilities of qualifying by system-level test, because there is only loss in observing components' performance in configurations that would never be used in the system.

4.4.1. Observability of Parameters in TID Tests

There is an advantage of TID tests over SEE tests in terms of observability coming from the fact that it is possible to perform remote testing after TID irradiation: DUT (device under test) or SUT might be removed from the beam area to perform electrical/functional characterization.

4.4.2. Observability of Parameters in SEE Tests

Contrary to TID tests, SEE test measurements need to be performed in-situ. This may lead to significant limitations due to facility mechanical constraints, such as length and/or number and type of available cable connections to monitor electrical and functional parameters (this limitation was met during each SEE test campaign of the PoL, but was particularly cumbersome during test campaigns at CHARM facility, see [6,24,25]).

5. Discussion

Regardless of the different limitations of the system-level testing that were presented in chapter 4, this approach might be still useful in product development. Particularly, let us further discuss the example of SiP modules from 3D PLUS: the company has a database of tests at the device level, so they have knowledge of the device sensitivity to radiation and lot-to-lot variations. They are then able to identify risky devices and build systems with expected initial radiation performance reasonably good. In this context, system-level radiation tests might be used to verify the performance of 3D PLUS products.

Moreover, the electronic circuits are created by the company designers, who are then able to apply their own proprietary mitigations technics. It is essential to take benefit of the test at the system level because they are able to identify the efficiency of the mitigation techniques and to identify what could be improved. As the SiP is a white box for the company designers, they can increase the number of observability points, to have better insight into system details, particularly during the prototype-level (board-level) testing.

Furthermore, at all design phases, it is possible to identify failure modes and error signatures for the system and to evaluate general TID performance. This might be an interesting added value in the design process.

Qualification limitations presented in chapter 4, might be at least partially addressed in the SiP product development flow:

- to define the worst case test conditions, testing in different configurations is needed, but the knowledge about the system design may facilitate defining the configurations to verify;
- defining the number of SUTs required to have good enough statistical results is important; testing the SiP models at early development phases (e.g., without full mechanical, thermal qualification) might significantly reduce the cost, if it turns out that a high number of SUTs is needed to test;
- in terms of beam limitations, which are particularly strong for SEE tests, testing the SiP at 2D board level (instead of the 3D module) would usually increase the value of performed tests, because the beam penetration might be lower, therefore the available LET range of particles increases;
- in terms of the system observability, limitations are again stronger for SEE tests, but here the solution could be the use of the supplementary instrumentation in digital SUTs [31] to increase the observability (for the cost of additional development).

Nevertheless, the complete system qualification, i.e., the one that would overcome all the qualification limitations discussed in chapter 4, is difficult. Furthermore, the cost of performing such complete qualification might be too high to make it a useful tool for the product or mission development. In some cases, simplification of the qualification is possible thanks to the knowledge about the system design or thanks to the knowledge about the target environment. However, for most of the cases, complete system qualification by system-level test remains difficult, and it is justified to consider the development of the limited qualification approach, that might be easier to apply, give valuable information, while bringing some level of risk that needs to be accepted.

Limited qualification scenarios and risk acceptance should be considered in the context of the system reliability and availability. High reliability space missions require maximum reliability (the system is not destroyed and is working within the specifications) and availability close to 100% (availability defined as a percentage of time that the system can be used during the mission, this percentage might be decreased by, e.g., power cycling of the system due to SEFI (single event functional interrupt)). High reliability missions will never accept qualification scenarios that introduce some risk in the reliability and will rarely accept any risk in the availability, even if the effort of the qualification will be lower.

On the other hand, for *New Space* missions, controllable risk acceptance is one of the fields to look for savings. Lower system availability, if justified, might be acceptable. Even slightly reduced system reliability might be considered if the system is understood as a single satellite in the constellation, where the mission (constellation) success is possible with partial loss of satellites. This opens the doors for use of limited TID/SEE qualification scenarios for the final qualification in *New Space* applications.

6. Conclusions

Based on the experimental results from several radiation test campaigns, the overview of opportunities and limitations of system qualification using the system-level radiation testing was presented. Impact on the SiP product development was discussed. Limitations for the SEE qualification are substantially stronger than for the TID qualification, however, there are still important question marks regarding also TID qualification: how to define the worst case conditions, how important is testing in these conditions, how many system units should be tested to have a high confidence level of results? Future works are planned to answer some of these questions, as well as to investigate the limited qualification approach for system qualification.

Author Contributions: Conceptualization, T.R., F.S. and P.-X.W.; methodology, T.R., F.S. and P.-X.W.; software, T.R.; validation, T.R., F.S. and P.-X.W.; formal analysis, T.R. and F.S.; investigation, T.R., F.S. and P.-X.W.; resources, T.R., F.S. and P.-X.W.; data curation, T.R.; writing—original draft preparation, T.R. and F.S.; writing—review and editing, T.R., F.S. and P.-X.W.; visualization, T.R.; supervision, F.S. and P.-X.W.; project administration, F.S. and P.-X.W.; funding acquisition, F.S. and P.-X.W. All authors have read and agreed to the published version of the manuscript.

Funding: Laser: X-rays and Co^{60} tests have been performed using the PRESERVE Platform which has been funded by the Occitanie Region and the EU via ERDF funds. This work has received funding from the European Union's Horizon 2020 research and innovation program under the Marie Skłodowska-Curie grant agreement number 721624.

Institutional Review Board Statement: Not applicable.

Informed Consent Statement: Not applicable.

Data Availability Statement: Most of data used for this study might be found at [6,22–25].

Acknowledgments: This work was realized within the RADSAGA project. The RADSAGA Innovative Training Network project has received funding from the European Union's Horizon 2020 research and innovation program under the Marie-Skłodowska-Curie grant agreement number 721624. Authors wish to thank colleagues from CERN for their support in organizing and performing tests at the CHARM facility: Rubén García Alía, Salvatore Danzeca, Chiara Cangialosi, Pablo Fernández-Martínez, Maria Kastriotou, Andrea Coronetti, Nourdine Kerboub and Corinna Martinella.

Conflicts of Interest: The authors declare no conflict of interest. The funders had no role in the design of the study; in the collection, analyses, or interpretation of data; in the writing of the manuscript, or in the decision to publish the results.

References

1. Mangeret, R. Radiation hardness assurance: How well assured do we need to be? In Proceedings of the 2018 IEEE NSREC Conference Short Course Notebook, Kona, HI, USA, 16–20 July 2018.
2. Campola, M.; Pellish, J. Radiation Hardness Assurance: Evolving for NewSpace. In Proceedings of the 2019 RADECS Short Course Notebook, Montpellier, France, 16–20 September 2019.

3. Perez, F. Overview of ESA CubeSat Missions and Radiation Testing on CubeSat Electronics. In Proceedings of the 2021 RADECS Conference Short Course Notebook, Vienna, Austria, 13 September 2021.
4. Coronetti, A.; Alía, R.G.; Budroweit, J.; Rajkowski, T.; Da Costa Lopes, I.; Niskanen, K.; Söderström, D.; Cazzaniga, C.; Ferraro, R.; Danzeca, S.; et al. Radiation hardness assurance through system-level testing: Risk acceptance, facility requirements, test methodology and data exploitation. *IEEE Trans. Nuclear Sci.* **2021**, *68*, 958–969. [CrossRef]
5. Hi-Rel Point-of-Load DC/DC Converter. 3D PLUS Product Datasheet, 3DFP-0296-REV. 6 April 2017.
6. Rajkowski, T.; Saigné, F.; Pouget, V.; Wrobel, F.; Touboul, A.; Boch, J.; Kohler, P.; Dubus, P.; Wang, P.X. Analysis of SET Propagation in a System in Package Point of Load Converter. *IEEE Trans. Nuclear Sci.* **2020**, *67*, 1494–1502. [CrossRef]
7. Vassal, M.-C.; Dubus, P.; Fiant, N. Electrical performances and radiation qualification tests results of a highly integrated and Space qualified Point of Load Converter. In Proceedings of the 12th European Conference on Radiation and Its Effects on Components and Systems, Sevilla, Spain, 19–23 September 2011.
8. Rousselet, M.; Adell, P.C.; Sheldon, D.J.; Boch, J.; Schone, H.; Saigné, F. Use and benefits of COTS board level testing for radiation hardness assurance. In Proceedings of the RADECS 2016, Bremen, Germany, 19–23 September 2016.
9. Fernández, G.; Bárcena, J.; Muñoz, E. TID test on ITER Interlock Discharge Loop Interface Box (DLIB) system, an example of radiation test at equipment level. In Proceedings of the RADECS 2017, Geneva, Switzerland, 2–6 October 2017.
10. Virmontois, C.; Belloir, J.; Beaumel, M.; Vriet, A.; Perrot, N.; Sellier, C.; Bezine, J.; Gambart, D.; Blain, D.; Garcia-Sanchez, E.; et al. Dose and Single-Event Effects on a Color CMOS Camera for Space Exploration. *IEEE Trans. Nuclear Sci.* **2019**, *66*, 104–110. [CrossRef]
11. O'Neill, P.M.; Badhwar, G.D.; Culpepper, W.X. Risk assessment for heavy ions of parts tested with protons. *IEEE Trans. Nuclear Sci.* **1997**, *44*, 2311–2314. [CrossRef] [PubMed]
12. Guertin, S. *Board Level Proton Testing Book of Knowledge for NASA Electronic Parts and Packaging Program*; NASA JPL Publication: Pasadena, CA, USA, 2017.
13. Guertin, S. *Lessons and Recommendations for Board-Level Testing with Proton*; Small Satellite Conference: Logan, UT, USA, 2018.
14. De Bibikoff, A.; Lamberbourg, P. Method for System-level testing of COTS electronic board under High Energy Heavy Ions. *IEEE Trans. Nuclear Sci.* **2020**, *67*, 2179–2187. [CrossRef]
15. Alía, R.G.; Martínez, P.F.; Kastriotou, M.; Brugger, M.; Bernhard, J.; Cecchetto, M.; Cerutti, F.; Charitonidis, N.; Danzeca, S.; Gatignon, L.; et al. Ultraenergetic heavy-ion beams in the CERN accelerator complex for radiation effects testing. *IEEE Trans. Nuclear Sci.* **2019**, *66*, 458–465. [CrossRef]
16. Fernández-Martínez, P.; Alía, R.G.; Cecchetto, M.; Kastriotou, M.; Kerboub, N.; Tali, M.; Wyrwoll, V.; Brugger, M.; Cangialosi, C.; Cerutti, F.; et al. SEE tests with ultra energetic Xe ion beam in the CHARM facility at CERN. *IEEE Trans. Nuclear Sci.* **2019**, *66*, 1523–1531. [CrossRef]
17. Mekki, J.; Brugger, M.; Alia, R.G.; Thornton, A.; Mota, N.C.D.S.; Danzeca, S. CHARM: A Mixed Field Facility at CERN for Radiation Tests in Ground, Atmospheric, Space and Accelerator Representative Environments. *IEEE Trans. Nuclear Sci.* **2016**, *63*, 2106–2114. [CrossRef]
18. Thornton, A. CHARM Facility Test Area Radiation Field Description. Available online: http://cds.cern.ch/record/2149417 (accessed on 14 January 2022).
19. Secondo, R.; Alia, R.G.; Peronnard, P.; Brugger, M.; Masi, A. System level radiation characterization of a 1U CubeSat based on CERN radiation monitoring technology. *IEEE Trans. Nuclear Sci.* **2018**, *65*, 1694–1699. [CrossRef]
20. Coronetti, A.; Manni, F.; Mekki, J. Mixed-field radiation qualification of a COTS space on-board computer along with its CMOS camera payload. In Proceedings of the RADECS 2019, Montpellier, France, 16–20 September 2019.
21. RADSAGA System-Level Testing Guideline for Space Systems. RADSAGA 2020. Available online: https://edms.cern.ch/document/2423146/2.1 (accessed on 14 January 2022).
22. Rajkowski, T.; Saigné, F.; Niskanen, K.; Boch, J.; Maraine, T.; Kohler, P.; Dubus, P.; Touboul, A.; Wang, P.-X. Comparison of the Total Ionizing Dose Sensitivity of a System in Package Point of Load Converter Using Both Component- and System-Level Test Approaches. *Electronics* **2021**, *10*, 1235. [CrossRef]
23. RADSAGA Test Report No. 01: GANIL RADSAGA Campaign, Date: 15/01/2019, Version no. 2.1—EDMS 2086126/1, RADSAGA 2019. Available online: https://radsaga.web.cern.ch/test-reports (accessed on 12 January 2022).
24. RADSAGA Test Report No. 02: CHARM RADSAGA Campaign #1, Date: 23/01/2019, Version no. 2.1—EDMS 2086128/1, RADSAGA 2019. Available online: https://radsaga.web.cern.ch/test-reports (accessed on 12 January 2022).
25. RADSAGA Test Report No. 05: CERN Heavy Ions RADSAGA Campaign, Date: 27/08/2019, Version no. 1.1—EDMS 2215703 v.1, RADSAGA 2019. Available online: https://radsaga.web.cern.ch/test-reports (accessed on 12 January 2022).
26. Single Event Effects Test Method and Guidelines. ESCC Basic Specification No. 25100, Issue 2, October 2014. Available online: http://escies.org/escc-specs/published/25100.pdf (accessed on 20 January 2022).
27. Total Dose Steady-State Irradiation Test Method. ESCC Basic Specification No. 22900, Issue 5, June 2016. Available online: http://escies.org/escc-specs/published/22900.pdf (accessed on 20 January 2022).
28. Schwank, J.R.; Shaneyfelt, M.R.; Dodd, P.E. Radiation Hardness Assurance Testing of Microelectronic Devices and Integrated Circuits: Radiation Environments, Physical Mechanisms, and Foundations for Hardness Assurance. *IEEE Trans. Nuclear Sci.* **2013**, *60*, 2074–2100. [CrossRef]

29. Schwank, J.R.; Shaneyfelt, M.R.; Dodd, P.E. Radiation Hardness Assurance Testing of Microelectronic Devices and Integrated Circuits: Test Guideline for Proton and Heavy Ion Single-Event Effects. *IEEE Trans. Nuclear Sci.* **2013**, *60*, 2101–2118. [CrossRef]
30. Söderström, D. Board and system level testing in other RADSAGA facilities and beyond. presentation during RADSAGA System Level Test Review, CERN, Geneva, Switzerland, 11–13 November 2019. Available online: https://indico.cern.ch/event/843292 /contributions/3541581/ (accessed on 20 January 2022).
31. Lopes, I.C.; Pouget, V.; Wrobel, F.; Saigne, F.; Touboul, A.; Røed, K. Development and evaluation of a flexible instrumentation layer for system-level testing of radiation effects. In Proceedings of the 2020 IEEE Latin-American Test Symposium (LATS), Maceio, Brazil, 30 March–2 April 2020.

 *electronics*

Article

TAISAM: A Transistor Array-Based Test Method for Characterizing Heavy Ion-Induced Sensitive Areas in Semiconductor Materials

Jinjin Shao, Ruiqiang Song *, Yaqing Chi, Bin Liang and Zhenyu Wu

College of Computer, National University of Defense Technology, Changsha 410073, China; shaojinjin308@163.com (J.S.); yqchi@nudt.edu.cn (Y.C.); liangbin110@126.com (B.L.); zxdwzy@foxmail.com (Z.W.)
* Correspondence: songrq07@nudt.edu.cn

Abstract: The heavy ion-induced sensitive area is an essential parameter for space application integrated circuits. Circuit Designers need it to evaluate and mitigate heavy ion-induced soft errors. However, it is hard to measure this parameter due to the lack of test structures and methods. In this paper, a test method called TAISAM was proposed to measure the heavy ion-induced sensitive area. TAISAM circuits were irradiated under the heavy ions. The measured sensitive areas are 1.75 μm^2 and 1.00 μm^2 with different LET values. TAISAM circuits are also used to investigate the layout structures that can affect the sensitive area. When the source region of the target transistor is floating, the heavy ion-induced sensitive area decreases by 28.5% for the target PMOS transistor while it increases by more than 28% for the target NMOS transistor. When the well contacts are added, the heavy ion-induced sensitive area decreases by more than 25% for the target PMOS transistor while it remains unchanged for the target NMOS transistor. Experimental results directly validate that the two structures significantly affect the heavy ion-induced sensitive area.

Keywords: radiation effect; radiation test method; heavy ion; sensitive area; parasitic bipolar amplification

Citation: Shao, J.; Song, R.; Chi, Y.; Liang, B.; Wu, Z. TAISAM: A Transistor Array-Based Test Method for Characterizing Heavy Ion-Induced Sensitive Areas in Semiconductor Materials. *Electronics* **2022**, *11*, 2043. https://doi.org/10.3390/electronics11132043

Academic Editor: Paul Leroux

Received: 6 June 2022
Accepted: 27 June 2022
Published: 29 June 2022

Publisher's Note: MDPI stays neutral with regard to jurisdictional claims in published maps and institutional affiliations.

1. Introduction

When a high energy incident particle strikes a semiconductordevice, it interacts with the semiconductor material and loses energy along its incident path [1,2]. The lost energy transfers to the semiconductor atoms and ionizes electron–hole pairs [3]. These ionized electron–hole pairs move into the entire semiconductor device due to drift and diffusion. Transistors collect the ionized electron–hole pairs, which produce unwanted transient currents in circuit nodes [4,5]. These transient currents propagate along the circuit path and produce soft errors according to the circuit responses [6]. They may result in serious consequences to the entire chip, system, or even a spacecraft [7]. The heavy ion-induced sensitive area is an essential parameter for space-application integrated circuits. It is useful to evaluate and mitigate the soft errors. For instance, the multiple-node charge collection effect significantly increases heavy ion-induced soft errors [8,9]. Increasing the distance between transistors in layout is a useful hardening method for mitigating soft errors induced by this effect [10–12]. However, due to the lack of heavy ion-induced sensitive area data, it is difficult for circuit designers to determine the required increased distance. Some previous works have reported standard cell-based test circuits to indirectly investigate the heavy ion-induced sensitive area [13,14]. However, it is hard to accurately obtain the sensitive area data due to the large layout area of standard cells. Therefore, it is necessary to propose novel test methods that can directly measure the heavy ion-indued sensitive area. In this paper, a special test method called Transistor Array-based Ion Sensitive Area Measurement (TAISAM) is presented to directly measure the heavy ion-induced sensitive area. The test principle is similar with the SRAM-based measurement [15]. A transistor array is used to quantify the transistors that are struck by heavy ions. Then, the heavy ion-induced sensitive area is calculated by multiplying the equivalent area of each transistor test chip that contains TAISAM circuits that were designed and fabricated by the commercial 65 nm bulk CMOS technology, irradiated under the heavy ions. The heavy ion-induced sensitive areas are reported and discussed based on the experimental results.

2. TAISAM Test Structure

The basic TAISAM cell topology is shown in Figure 1. It consists of a conventional D flip-flop and a target transistor (PMOS or NMOS). The gate of the target transistor is fixed to a constant voltage to switch off the transistor. The gate voltage is connected to power for target PMOS transistor while it is connected to the ground for a target NMOS transistor. The drain region of the target transistor is connected to the D flip-flop, and the source region of the target transistor is connected to power or ground. Due to the special connection of test circuits, the stored value of D flip-flops determines the heavy ion sensitivity of target transistors. For instance, when the stored value of D flip-flops is zero, the drain voltage of target PMOS transistor is also zero. The drain voltage is lower than the gate/source voltage and the ionized electron–hole pairs can be collected by target PMOS transistor. Therefore, the target PMOS transistors are sensitive to heavy ions. Otherwise, when the stored value of D flip-flops is one, the drain voltage of target PMOS transistor is also one. The drain voltage is equal to the gate/source voltage and the ionized electron–hole pairs cannot be collected by PMOS transistor; thus, the target PMOS transistors are insensitive to heavy ions.

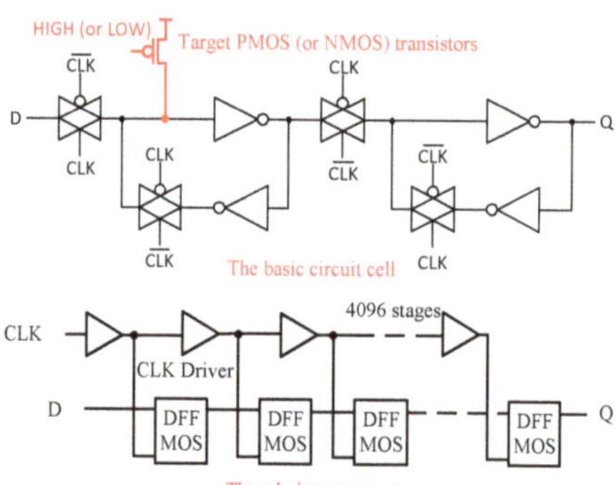

Figure 1. The basic TAISAM cell contains a flip-flop and a target transistor. Several TAISAM cells consist of an entire test circuit.

Four thousand and ninety-six stage basic cells are used to constitute a shift register circuit. All target transistors are placed uniformly and side-by-side to constitute a 256×16 transistor arrays, as shown in Figure 2. The target transistor arrays and the D flip-flops are placed separately to make sure a heavy ion only strikes the transistor arrays or the D flip-flops each time. If a heavy ion strikes the center region of the transistor arrays, transistors inside the heavy ion-induced sensitive area collect the ionized electron–hole pairs and produce transient currents at the storage node of D flip-flops simultaneously. If the number of collected electron–hole pairs is larger than the critical charge of D flip-flops, the initial value would be changed by transient currents. It causes single-event upset (SEU) effect. Then, the changed values shift to the outputs of the circuit. The number of transistors which inside the heavy ion-induced sensitive area is determined by counting the changed values on the output port of the test circuit. Based on the number of SEU, the heavy ion-induced sensitive area is calculated by the following equation:

$$S_{ion} = N_{\max} \times S_{transistor} \tag{1}$$

where $N_{\max}$ is the maximum number of SEU. $S_{transistor}$ is the equivalent area of each transistor. Note that both D flip-flops and target PMOS transistors are sensitive to heavy ions. It may cause SEUs when a heavy ion strikes the D flip-flop layout or the transistor array layout. Due to the large layout area of one D flip-flop, the number of SEUs is lower than that when a heavy ion hits the transistor array. When a heavy ion strikes the transistor array, there are two situations. If the heavy ion strikes the edge of transistor array, it does not fully reveal the heavy ion-indued sensitive area. The number of SEU is smaller than when a heavy ion strikes the center of the transistor array. Therefore, the maximum number of SEU represents a heavy ion hit the center of the transistor array. The heavy ion-induced sensitive area is calculated by the maximum number of SEU.

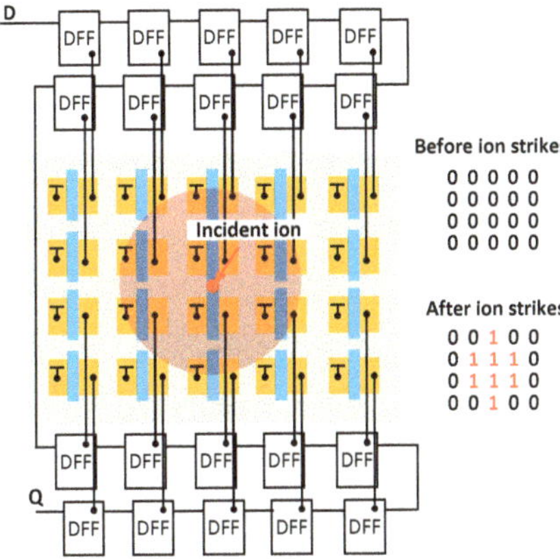

Figure 2. The layout schematic of a TAISAM circuit. The transistor array is placed in the center region and the flip-flops are placed beside the transistor array.

In addition to obtaining a statistical value of the heavy ion-induced sensitive area, TAISAM is also able to draw planar maps of the heavy ion-induced sensitive area. For instance, the diameter of the heavy ion-induced sensitive area in the horizontal and vertical directions can be evaluated by the output data after one ion striking, as shown in Figure 3. Moreover, TAISAM can also map the heavy ion-induced sensitive areas with different critical charge values. Since the target transistor-induced SEU is dependent on the critical charge of the D flip-flop, it can adjust the critical charge values to obtain different heavy ion-induced sensitive areas. According to these measured results, the entire heavy ion-induced sensitive area with different charge values can be determined.

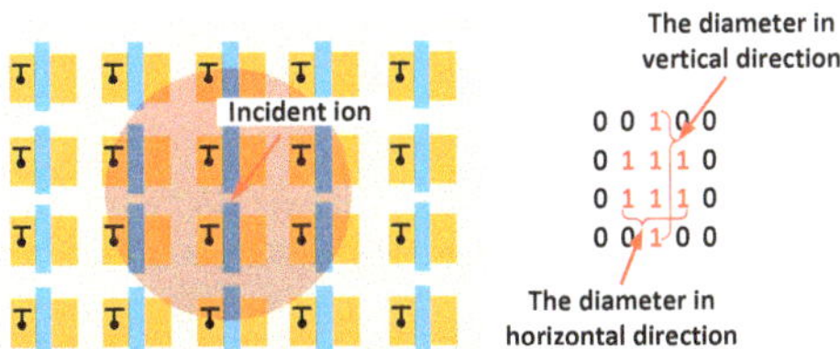

Figure 3. TAISAM can be used to determine the diameter of the ion-induced sensitive area.

3. Test Chip Design and Experimental Setup

A test chip layout was designed by the commercial 65 nm bulk CMOS technology. It consisted of six TAISAM circuits (A–F) with different transistor arrays. The detailed descriptions of these test circuits are shown in Table 1. Test circuits A and B consisted of normal transistor arrays, as shown in Figure 4. The source region is connected to power for target PMOS transistors in test circuit A and it is connected to ground for target NMOS transistors in test circuit B. Test circuit C and D consisted of transistor arrays with the floating source structure. The source region of target transistors is not connected to power or ground. For PMOS transistors, the floating source structure breaks the parasitic bipolar conduction of drain-well source [16]. This structure mitigates the parasitic bipolar amplification (PBA) effect and may significantly affect the ion-induced sensitive area. For NMOS transistors, the source region could not help collect ionized electron–hole pairs because of the floating structure. Therefore, it may also affect the ion-induced sensitive area. N-well or P-well contacts are, respectively, added near the edge of transistor arrays in test circuit E and F. They are used to investigate the influence of well contacts on the heavy ion-induced sensitive area.

Table 1. Transistor array types in the test chip.

No.	Transistor Array Description
A	Normal PMOS array
B	Normal NMOS array
C	PMOS array (Source float)
D	NMOS array (Source float)
E	PMOS array (Well contact)
F	NMOS array (Well contact)

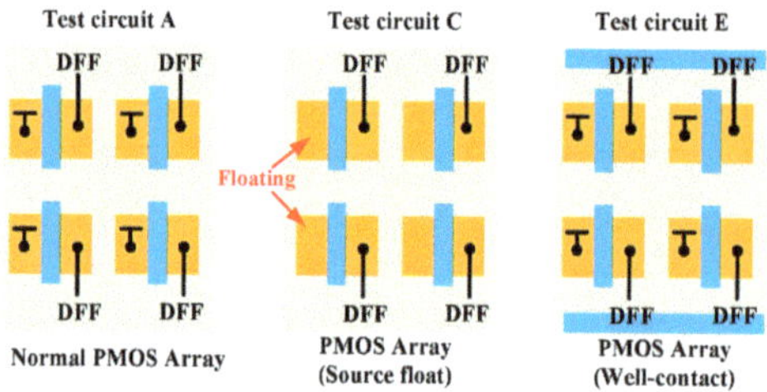

Figure 4. The schematic of the different transistor arrays.

The detailed test chip layout is shown in Figure 5. The test chip size is 0.9 mm height and 0.6 mm width. The width and length of the target transistor sizes are 0.12 μm and 0.06 μm, respectively. The distance between two target transistors is 0.13 μm, which is the minimum value of the layout spacing rule. The height of a transistor array in the vertical direction is 3.87 μm and it is 264.4 μm in the horizontal direction. Based on our previous experimental results, heavy ions impacted no more than three inverters when the LET is smaller than 40 MeV·cm^2/mg [13]. The calculated diameter of heavy ions is no more than 2 μm. Therefore, the size of the transistor array is enough to measure the heavy ion-induced sensitive area. The layout area of a transistor array is 1024 μm^2 and the equivalent area of each target transistor is 0.25 μm^2. It is worth to note that the layout of the target transistor array is placed in the central area while the layout of the D flip-flop is placed above and below the target transistor array. The target transistor arrays and the D flip-flops in the layout are separated to ensure a heavy ion only strikes one of them. The D flip-flop with PMOS/NMOS balanced X1 driven strength is used to connect the target transistor, since it had the minimum critical charge value to made an exact measurement.

Heavy ion experiments were conducted at the HI-13 Tandem Accelerator in China Institute of Atomic Energy and the Heavy Ion Research Facility in Lanzhou (HIRFL) cyclotrons in Institute of Modern Physics, Chinese Academy of Sciences. Three heavy ions with different parameters were used, as shown in Table 2. The heavy ion dose rate was 1×10^4 ions/(cm^2·s), which was determined by the test chip area and the operation frequency. This dose rate value makes sure that the data can be shifted to the output ports before the next heavy ion struck the test circuit. The fluence rate was 1×10^7 ions/cm^2 to make sure enough measurement data are obtained. The number of heavy ions is a factor to impact TAISAM measurement results. TAISAM circuits are used to investigate the sensitive area when one heavy ion hit the circuits. During the heavy ion experiment, it could not control the heavy ion incident locations. If the number of heavy ions is small, heavy ions may not hit the center of transistor arrays and the measurement results could not fully reveal the ion-induced sensitive area. When the number of heavy ions is large, it increases the probability that heavy ions hit the center of transistor arrays. The measurement results can fully reveal the ion-induced sensitive area. The entire test system consisted of a test chip and other necessary chips, such as field programmable gate arrays (FPGAs) and serial communication chips. The test chip's operation frequency is 40MHz and the serial interface's baud rate is 115,200 bps. FPGAs connected all input and clock ports of the test chip to provide input and clock signals. They were also used to capture output signals and count SEUs when the test chip was irradiated. By conducting these heavy ion experiments, the error counts were exported to the computer by the serial communication interface.

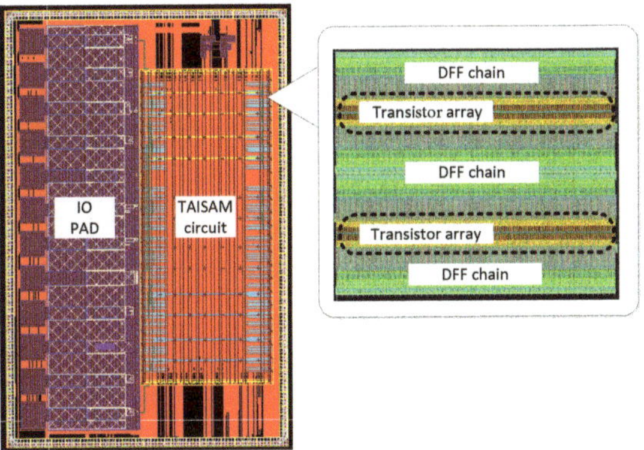

Figure 5. The detailed layout placement in the test chip.

Table 2. Heavy ions used in the experiment.

Ion	Energy at the Silicon Surface (MeV)	Effective LET (MeV·cm^2/mg)	Range (um)
Cl	165	15.2	51.8
Ge	205	37.6	35.5
Kr	835.5	39.8	41.2

The test chip layout was fabricated by the commercial 65 nm bulk CMOS technology. To avoid experimental result variations caused by the sample, five test chips were packaged and were irradiated in the same dynamic test mode. In previous studies, the dynamic test mode had been widely used to estimate the SEU sensitivity of circuits [17–20]. In this paper, TAISAM circuits were irradiated with constant 0 data or constant 1 data. Since the input data were fixed, soft errors owing to the clock tree were avoided. The transient response in the clock tree may cause the data to shift forward without any error. Therefore, SEU only occurred when a heavy ion struck target transistors or D flip-flops.

4. Experimental Results and Discussion

By comparing the experimental results of each test chip, the number of SEUs shows the same distributions. Figure 6 shows one test chip's experimental result with different LET values. The statistical SEU distributions show obvious discrepancies with different data patterns. When the input data are a constant 0, both PMOS transistors and D flip-flops are sensitive to heavy ions for test circuit A. When the input data are a constant 1, only D flip-flops are sensitive to heavy ions for test circuit A. Similarly, SEU distributions are observed for test circuit B. The max number of SEU also shows a difference between the two different input data.

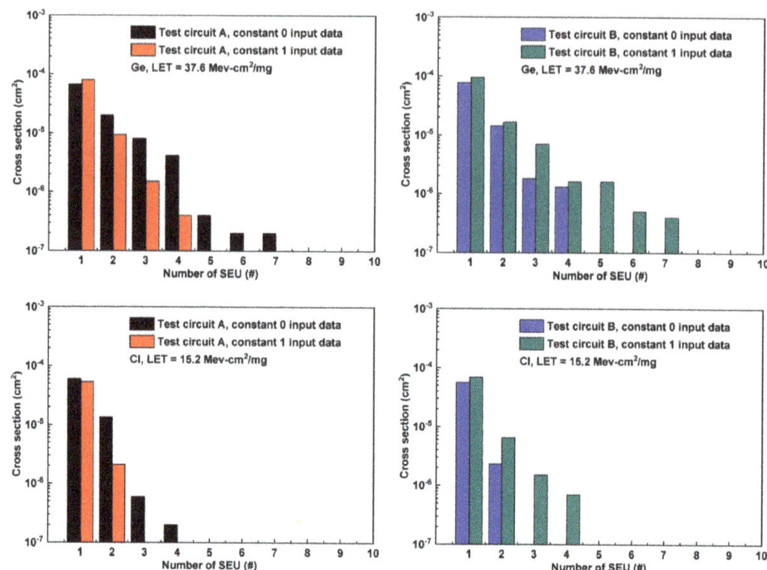

Figure 6. Measured results with different LET values. Test circuit A and B have the normal connections of target transistors.

For test circuit A, when both target transistors and D flip-flops are sensitive to heavy ions, the maximum number of SEU is 7. When only D flip-flops are sensitive to heavy ions, the maximum number of SEU is 4. Measured results confirm that the number of SEU 5, 6, and 7 are obtained when a heavy ion strikes target transistors. The number of SEU 5 and 6 may occur when a heavy ion strikes the edge region of the transistor arrays. However, it does not fully reveal the heavy ion-induced sensitive area. Therefore, the heavy ion-induced sensitive area can be determined by the maximum number of SEU. According to the Equation (1), the measured heavy ion-induced sensitive areas are 1.75 μm^2 and 1.00 μm^2 when the LET values are 37.6 MeV·cm^2/mg and 15.2 MeV·cm^2/mg, respectively.

Figure 7 shows the measured heavy ion-induced sensitive areas of test circuits A–D. For PMOS transistors, the source and drain region are P type doping. The N-well region is N type doping. The source-well drain constitutes a parasitic PNP junction transistor. When a heavy ion hit PMOS transistors, the ionized electron–hole pairs not only diffuse to the PMOS transistors but also lead to the collapse of the N-well potential. The lower N-well potential activates the parasitic PNP junction transistor. It results in the charge injection from the source region to the drain region and causes the PBA effect. Therefore, although the ionized electron–hole pairs may not diffuse to the PMOS transistors, they may also be impacted by heavy ions. The source floating structure breaks the parasitic PNP junction transistor. In our experimental results, the floating source structure hardly affects the heavy ion-induced sensitive area at low LET value. However, it significantly reduces the heavy ion-induced sensitive area at high LET value. The PBA effect slightly causes the injection of holes into the N-well at lower LET value due to the perturbation of N-well voltage is not obvious [21–23]. The PBA effect does not increase the density of the ionized electron–hole pairs at low LET value and the heavy ion-induced sensitive area does not change at low LET values.

However, the perturbation of N-well voltage results in the injection of holes into the N-well at high LET value. The injected holes increase the density of the ionized electron–hole pairs and affect more transistors. Since the floating source structure of target PMOS transistors mitigates the PBA effect, the measured heavy ion-induced sensitive area in test circuit C decreases by 28.5% compared with the data of test circuit A. For NMOS transistors, the floating source structure significantly affects the heavy ion-induced sensitive area at both low and high LET values. The source region of NMOS transistors could not help collect the ionized electron–hole pairs because of the floating structure [24]. The ionized electron–hole pairs can simultaneously affect more NMOS transistors by drifting and diffusing. Therefore, the floating source structure increases the heavy ion-induced sensitive area. The measured results in test circuit D increases by 28.5% and 50% compared with the data of test circuit B.

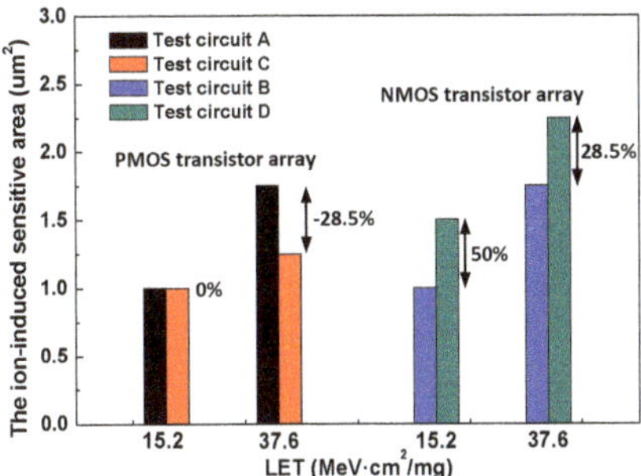

Figure 7. Effect of floating source regions on the measured ion-induced sensitive area with different LET values. Test circuit A and B have the normal connections of target transistors. The source regions of target transistors in test circuit C and D are floating.

Figure 8 shows the measured heavy ion-induced sensitive areas of test circuit A, B, E, and F. For PMOS transistors, the heavy ion-induced sensitive area significantly decreases by 25% and 28.5% with different LET values. The additional N-well contacts not only mitigate PBA effect but also help collect the ionized electron–hole pairs in the N-well. Thus, the measured results in test circuit E are smaller than that in test circuit A.

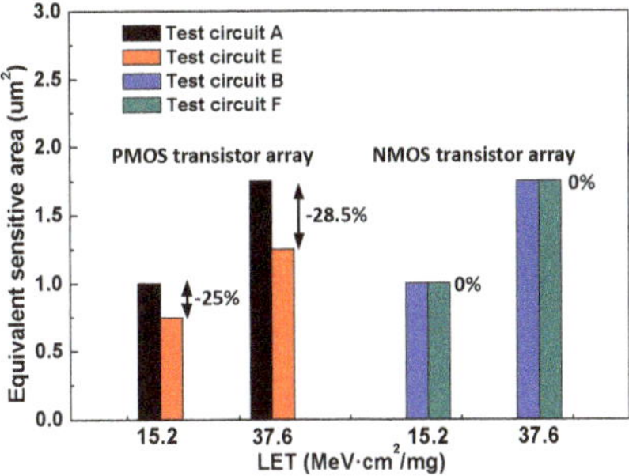

Figure 8. Effect of well contacts on the measured ion-induced sensitive area with different LET values. Test circuit A and B have the normal connections of target transistors. The well connections are added beside the target transistors in test circuit E and F.

However, the additional P-well contact has no effect for the heavy ion-induced sensitive area at both low and high LET values. Different from the N-well contact, the additional P-well contact slightly affects the ionized electron-hole pairs due to the ionized electrons and holes can spread through the entire substrate [25]. Therefore, although the P-well contact was added in test circuit F, the measured result is same as that in test circuit B. Experimental results directly confirm that only the N-well contact can affect the heavy ion-induced sensitive area.

In previous works, heavy ions with different types and energies may produce different sensitivity to circuits even if they have the same LET value [26]. One potential mechanism is that the tracks of ions in semiconductor materials are different. When heavy ions pass through semiconductor materials, some ion tracks prefer straight lines while other ion tracks prefer curves. Although the LET values of incident ions are same, the number of ionized electron–hole pairs may be different and they may produce different sensitivities to circuits. In this paper, TAISAM was also irradiated with ions that have the same LET but different types and energies. The measured results are shown in Figure 9. The maximum number of SEU is not changed even though the incident heavy ions are different. Thus, the calculated heavy ion-induced sensitive area has no change. One reason is that these ions have the similar tracks when they pass through semiconductor materials. Experimental results indicate that the heavy ion-induced sensitive area is only dependent on the LET values for some ions.

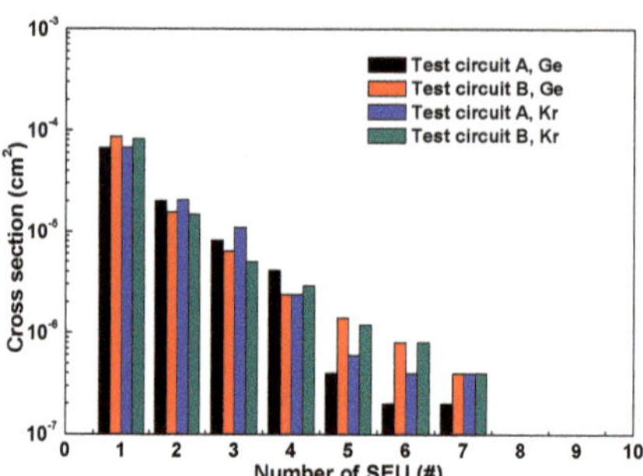

Figure 9. Measured results with different ion energies and species. The incident heavy ions have the similar LET values. However, they have different incident range and energies at the silicon surface.

5. Conclusions

The heavy ion-induced sensitive area is an important parameter for space application integrated circuits. It is essential to characterize it experimentally. This paper has presented a transistor array-based test method called TAISAM, which is used to investigate the heavy ion-induced sensitive area. TAISAM circuits were irradiated under heavy ions. Experimental results were reported, and the heavy ion-induced sensitive area was calculated. The source floating structure and the additional N-well connections are better for PMOS transistors in terms of decreasing the sensitive area because they mitigate the parasitic bipolar amplification effect. However, the source floating structure is worse for NMOS transistors because it does not help collect ionized electron–hole pairs. The additional P-well connection has no effect for the heavy ion-induced sensitive area. Experimental results are consistent with mechanism analyses.

Author Contributions: Conceptualization, J.S.; methodology, Y.C.; validation, B.L. and Z.W.; writing—original draft, J.S.; writing—review and editing, R.S. All authors have read and agreed to the published version of the manuscript.

Funding: This research was funded by the National University of Defense Technology research project (Grant No. ZK20-11).

Data Availability Statement: The datasets used and/or analyzed during the current study are available from the corresponding author upon reasonable request.

Acknowledgments: We wish to acknowledge the HI-13 team and the HIRFL team for heavy ion experiment supports.

Conflicts of Interest: The authors declare no conflict of interest.

Abbreviations

The following abbreviations are used in this manuscript:

TAISAM	Transistor Array-based Ion Sensitive Area Measurement
LET	linear energy transfer
SEU	single-event upset
PBA	parasitic bipolar amplification
FPGAs	field programmable gate arrays

References

1. Zhang, Z.; Arehart, A.; Cinkilic, E.; Chen, J.; Zhang, E.; Fleetwood, D.; Schrimpf, R.; McSkimming, B.; Speck, J.; Ringel, S. Impact of proton irradiation on deep level states in n-GaN. *Appl. Phys. Lett.* **2013**, *103*, 042102. [CrossRef]
2. Fleetwood, D.; Shaneyfelt, M.; Schwank, J. Estimating oxide-trap, interface-trap, and border-trap charge densities in metal-oxide-semiconductor transistors. *Appl. Phys. Lett.* **1994**, *64*, 1965–1967. [CrossRef]
3. Jun, B.; White, Y.; Schrimpf, R.; Fleetwood, D.; Brunier, F.; Bresson, N.; Cristoloveanu, S.; Tolk, N. Characterization of multiple Si/SiO_2 interfaces in silicon-on-insulator materials via second-harmonic generation. *Appl. Phys. Lett.* **2004**, *85*, 3095–3097. [CrossRef]
4. Ferlet-Cavrois, V.; Massengill, L.W.; Gouker, P. Single event transients in digital CMOS—A review. *IEEE Trans. Nucl. Sci.* **2013**, *60*, 1767–1790. [CrossRef]
5. Liu, J.; Sun, Q.; Liang, B.; Chen, J.; Chi, Y.; Guo, Y. Bulk bias as an analog single-event transient mitigation technique with negligible penalty. *Electronics* **2019**, *9*, 27. [CrossRef]
6. Azimi, S.; De Sio, C.; Rizzieri, D.; Sterpone, L. Analysis of Single Event Effects on Embedded Processor. *Electronics* **2021**, *10*, 3160. [CrossRef]
7. Tang, D.; He, C.; Li, Y.; Zang, H.; Xiong, C.; Zhang, J. Soft error reliability in advanced CMOS technologies-trends and challenges. *Sci. China Technol. Sci.* **2014**, *57*, 1846–1857. [CrossRef]
8. Black, J.D.; Dodd, P.E.; Warren, K.M. Physics of multiple-node charge collection and impacts on single-event characterization and soft error rate prediction. *IEEE Trans. Nucl. Sci.* **2013**, *60*, 1836–1851. [CrossRef]
9. Massengill, L.W.; Bhuva, B.L.; Holman, W.T.; Alles, M.L.; Loveless, T.D. Technology scaling and soft error reliability. In Proceedings of the 2012 IEEE International Reliability Physics Symposium (IRPS), Anaheim, CA, USA, 15–19 April 2012; p. 3C-1. [CrossRef]
10. Song, R.; Shao, J.; Liang, B.; Chi, Y.; Chen, J. *MSIFF: A Radiation-Hardened Flip-Flop via Interleaving Master-Slave Stage Layout Topology*; The Institute of Electronics, Information and Communication Engineers: Tokyo, Japan, 2020; p. 17-20190708. [CrossRef]
11. Jiang, S.; Liu, S.; Zheng, H.; Wang, L.; Li, T. Novel Radiation-Hardened High-Speed DFF Design Based on Redundant Filter and Typical Application Analysis. *Electronics* **2022**, *11*, 1302. [CrossRef]
12. Chi, Y.; Cai, C.; He, Z.; Wu, Z.; Fang, Y.; Chen, J.; Liang, B. SEU Tolerance Efficiency of Multiple Layout-Hardened 28 nm DICE D Flip-Flops. *Electronics* **2022**, *11*, 972. [CrossRef]
13. Huang, P.; Chen, S.; Chen, J.; Liang, B.; Chi, Y. Heavy-ion-induced charge sharing measurement with a novel uniform vertical inverter chains (UniVIC) SEMT test structure. *IEEE Trans. Nucl. Sci.* **2015**, *62*, 3330–3338. [CrossRef]
14. Huang, P.; Chen, S.; Chen, J.; Liang, B.; Song, R. The Separation Measurement of *P*-Hit and *N*-Hit Charge Sharing With an "S-Like" Inverter Chains Test Structure. *IEEE Trans. Nucl. Sci.* **2017**, *64*, 1029–1036. [CrossRef]
15. Casse, G.; Massari, N.; Franks, M.; Parmesan, L. A novel concept for a fully digital particle detector. *J. Instrum.* **2022**, *17*, P04010. [CrossRef]
16. Zhang, K.; Yamamoto, R.; Furuta, J.; Kobayashi, K.; Onodera, H. Parasitic bipolar effects on soft errors to prevent simultaneous flips of redundant flip-flops. In Proceedings of the 2012 IEEE International Reliability Physics Symposium (IRPS), Anaheim, CA, USA, 15–19 April 2012; p. 5B-2. [CrossRef]
17. He, Y.; Chen, S. Comparison of heavy-ion induced SEU for D-and TMR-flip-flop designs in 65-nm bulk CMOS technology. *Sci. China Inf. Sci.* **2014**, *57*, 1–7. [CrossRef]
18. Loveless, T.; Jagannathan, S.; Reece, T.; Chetia, J.; Bhuva, B.; McCurdy, M.; Massengill, L.; Wen, S.J.; Wong, R.; Rennie, D. Neutron-and proton-induced single event upsets for D-and DICE-flip/flop designs at a 40 nm technology node. *IEEE Trans. Nucl. Sci.* **2011**, *58*, 1008–1014. [CrossRef]
19. Warren, K.M.; Sternberg, A.L.; Black, J.D.; Weller, R.A.; Reed, R.A.; Mendenhall, M.H.; Schrimpf, R.D.; Massengill, L.W. Heavy ion testing and single event upset rate prediction considerations for a DICE flip-flop. *IEEE Trans. Nucl. Sci.* **2009**, *56*, 3130–3137. [CrossRef]
20. Song, R.; Shao, J.; Liang, B.; Chi, Y.; Chen, J. A Single-Event Upset Evaluation Approach Using Ion-Induced Sensitive Area. In Proceedings of the 2019 IEEE 13th International Conference on ASIC (ASICON), Chongqing, China, 29 October–1 November 2019; pp. 1–4. [CrossRef]
21. Jagannathan, S.; Gadlage, M.J.; Bhuva, B.L.; Schrimpf, R.D.; Narasimham, B.; Chetia, J.; Ahlbin, J.R.; Massengill, L.W. Independent measurement of SET pulse widths from N-hits and P-hits in 65-nm CMOS. *IEEE Trans. Nucl. Sci.* **2010**, *57*, 3386–3391. [CrossRef]
22. Ruiqiang, S.; Shuming, C.; Yankang, D.; Pengcheng, H.; Jianjun, C.; Yaqing, C. PABAM: A physics-based analytical model to estimate bipolar amplification effect induced collected charge at circuit level. *IEEE Trans. Device Mater. Reliab.* **2015**, *15*, 595–603. [CrossRef]

23. He, Y.B.; Chen, S.M. Experimental verification of the parasitic bipolar amplification effect in PMOS single event transients. *Chin. Phys. B* **2014**, *23*, 079401. [CrossRef]
24. Chen, J.; Chen, S.; He, Y.; Chi, Y.; Qin, J.; Liang, B.; Liu, B. Novel layout technique for N-hit single-event transient mitigation via source-extension. *IEEE Trans. Nucl. Sci.* **2012**, *59*, 2859–2866. [CrossRef]
25. Du, Y.; Chen, S.; Liu, B.; Liang, B. Effect of p-well contact on n-well potential modulation in a 90 nm bulk technology. *Sci. China Technol. Sci.* **2012**, *55*, 1001–1006. [CrossRef]
26. Reed, R.; Weller, R.; Mendenhall, M.; Lauenstein, J.M.; Warren, K.; Pellish, J.; Schrimpf, R.; Sierawski, B.; Massengill, L.; Dodd, P.; et al. Impact of ion energy and species on single event effects analysis. *IEEE Trans. Nucl. Sci.* **2007**, *54*, 2312–2321. [CrossRef]

Article

Comparison of Total Ionizing Dose Effects in 22-nm and 28-nm FD SOI Technologies

Zongru Li *, Christopher Jarrett Elash, Chen Jin, Li Chen, Jiesi Xing, Zhiwu Yang and Shuting Shi

Department of Electrical & Computer Engineering, University of Saskatchewan, Saskatoon, SK S7N 5A9, Canada; chris.elash@usask.ca (C.J.E.); chj482@usask.ca (C.J.); li.chen@usask.ca (L.C.); jiesi.xing@usask.ca (J.X.); zhy606@usask.ca (Z.Y.); shs085@usask.ca (S.S.)
* Correspondence: zol668@usask.ca

Abstract: Total ionizing dose (TID) effects from Co-60 gamma ray and heavy ion irradiation were studied at the 22-nm FD SOI technology node and compared with the testing results from the 28-nm FD SOI technology. Ring oscillators (RO) designed with inverters, NAND2, and NOR2 gates were used to observe the output frequency drift and current draw. Experimental results show a noticeable increased device current draw and decreases in RO frequencies where NOR2 ROs have the most degradation. As well, the functionality of a 256 kb SRAM block and shift-register chains were evaluated during C0-60 irradiation. SRAM functionality deteriorated at 325 krad(Si) of the total dosage, while the FF chains remained functional up to 1 Mrad(Si). Overall, the 22-nm FD SOI results show better resilience to TID effects compared to the 28-nm FD SOI technology node.

Keywords: 22-nm FD SOI; 28-nm FD SOI; Co-60; flip-flop (FF); heavy ion; radiation effects; ring oscillator (RO); static random-access memory (SRAM); total ionizing dose (TID)

Citation: Li, Z.; Elash, C.J.; Jin, C.; Chen, L.; Xing, J.; Yang, Z.; Shi, S. Comparison of Total Ionizing Dose Effects in 22-nm and 28-nm FD SOI Technologies. *Electronics* **2022**, *11*, 1757. https://doi.org/10.3390/electronics11111757

Academic Editor: Paul Leroux

Received: 1 May 2022
Accepted: 31 May 2022
Published: 1 June 2022

Publisher's Note: MDPI stays neutral with regard to jurisdictional claims in published maps and institutional affiliations.

1. Introduction

Integrated circuits (ICs) can experience various functionality issues when subjected to prolonged doses of radiation. These effects are often a reliability concern in long-term space missions where devices can spend years exposed to constant sources of radiation. Total ionizing dose (TID) effects in ICs can result in changes to gate propagation delays, leakage currents, and even loss of device functionality [1]. When ICs are exposed to ionizing radiation, positive charges are accumulated within the gate oxide and field oxide layers, thus resulting in less gate control of the device and an increased leakage current [2]. For PMOS transistors, the result of these charges can result in the device failing to turn on, whereas NMOS transistors become difficult to turn off [1,2]. The mechanisms of TID effects in bulk technologies are often simpler due to the inclusion of only one gate oxide layer; however, fully depleted silicon on insulator (FD SOI) technologies feature a more complex response to TID effects [3].

Transistors in the latest FD SOI technologies are fabricated on a very thin silicon (Si) layer over a buried oxide layer (BOX). With each transistor fabricated on a Si Island and isolated from other transistors by the BOX layer, the volume of active Si is minimal for each transistor. This allows for superior gate control over the channel region while reducing nodal capacitances, which yields faster logic gate switching times over those for bulk technologies. However, because FD SOI technologies have an additional parasitic structure due to the BOX layer, effects due to TID are more complex than bulk devices [4–9]. The BOX layer introduces a two-dimensional coupling effect between the front and back interfaces of the channel. This doubled coupling becomes a critical contribution to the ionizing dose response of FD SOI devices. In this case, FD SOI technologies tend to be more sensitive to TID than their bulk counterparts [10,11]. This is important to note due to the attractiveness of FD SOI technologies for use in space missions because of the technologies' inherent resilience to particle-induced single event effects (SEEs) [3,9]. There are already some experimental results from previous research on TID effects in the 28-nm FD SOI technology node [2,9]. These results show that a TID-induced gate delay increased significantly in that technology node, as well as leakage currents which impose barriers for it to be used for some space applications where tolerance to high total absorbed dose levels is required. Therefore, it is essential to investigate TID effects in the 22-nm FD SOI technology node and compare them with the results from the 28-nm FD SOI technology node.

To evaluate TID effects, a 22-nm FD SOI test chip was designed and fabricated. Ring oscillators (ROs), flip-flop (FF) chains, and a static random access memory (SRAM) block were chosen as the testing vehicles for evaluating the technology's susceptibility to TID effects; including both Co-60 and heavy ions irradiation sources [12]. RO circuits can offer insights into gate delays due to changes in the frequency of the circuits, as well as changes in power consumption as dosage increases. The FF chains and the SRAM block can be used to offer a broad perspective on device functionality. A similar test chip with 28-nm FD SOI technology was also fabricated previously, and the results from that test chip will be used in this paper for comparison purposes between the two technologies [9].

The organization of the rest of this paper is as follows. In Section 2, details of the test circuits are introduced. In Section 3, details of the test chip design and experimental setup are described. In Section 4, the experimental results will be presented and discussed. Lastly, conclusions are drawn in Section 5.

2. Description of Test Circuits

2.1. Ring Oscillators

Individual transistors are the ideal test vehicles for evaluating TID effects; however, there are many challenges regarding fabrication and testing. Additional design procedures are necessary to avoid the antenna effects during fabrication, leading to additional measurement errors. As explained by [13], measuring the change in the delay of a single transistor is difficult in practice, and can severely skew the results if the transistor has large variations in fabrication. Instead, ring oscillator (RO) circuits use the average parameter values of all transistors in the circuit, and as such, are well suited for the TID characterization of circuit-level parameters such as gate delay and power usage [8,14,15]. The use of averaged measurements allows for averaging statistical variations between individual gate parameters, thus ensuring that statistical variations between individual transistors in the RO circuit are incorporated into the overall RO characteristics [8]. Additionally, since ROs are often used for clock generation sources, it is important to understand how their performance can change as the dosage of ionizing radiation increases.

RO circuits are designed using an odd number of logic gates connected in a loop, which results in an unstable circuit oscillating at a fixed frequency. The oscillating frequency depends on the number of stages in the loop. The RO frequency is determined by the following equation:

$$f = \frac{1}{2 \times n \times t} \tag{1}$$

where f is the RO frequency, n is the number of stages in the RO loop, and t is the average delay of each RO stage. The equation for a single inverter delay is given below:

$$Delay_{inv} = \frac{C_L \times V_{DD}}{\frac{W}{L} \times \mu \times C_{ox} \times (V_{DD} - V_{th})^2} \tag{2}$$

where C_L is the load capacitance, V_{DD} is the supply voltage, W and L are the width and length of the gate, μ is the carrier mobility, C_{ox} is the capacitance of the oxide layer, and V_{th} is the gate threshold voltage [16]. TID effects will alter both the mobility and the threshold of the device, which will thus result in a measurable difference in the delay of the gate, and therefore the frequency of the oscillator.

A variety of different ROs were constructed with 2-input NAND (NAND2) gates, 2-input NOR (NOR2) gates, and inverters, as shown in Figure 1. For each type of gate, multiple ROs were designed, each with a different number of gates to achieve different oscillating frequencies. It should be noted that they were designed based on the cells from the standard cell library without any special layout design techniques. The inverter ROs included four different frequency options, and the NAND2 ROs and NOR2 ROs both had three different frequency options. Tables 1–3 show the ROs' expected oscillation frequencies normalized to their respective fastest oscillation frequency as measured by post-layout simulations from the foundry supplied process design kit (PDK).

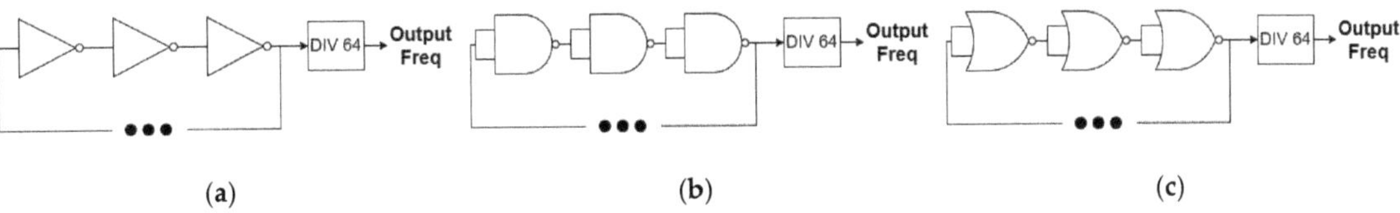

Figure 1. Three different RO designs: (**a**) Inverter-based RO design; (**b**) NAND2-based RO design; (**c**) NOR2-based RO design.

Table 1. Inverter-based programmable RO simulation results.

Delay Block	Number of Delay Stages	Output Frequency (n.u.)
S0	21	1.00
S1	29	0.77
S2	45	0.57
S3	69	0.36

Table 2. NAND-based programmable RO simulation results.

Delay Block	Number of Delay Stages	Output Frequency (n.u.)
S0	13	1.00
S1	23	0.68
S2	49	0.39

Table 3. NOR-based programmable RO simulation results.

Delay Block	Number of Delay Stages	Output Frequency (n.u.)
S0	13	1.00
S1	23	0.68
S2	49	0.38

The output of oscillators for each respective gate was connected to a multiplexer and a frequency divider circuit, as shown in Figure 2. The inclusion of the divider was needed as the high frequencies of the oscillators were not able to be directly captured by the chip's IO pads, which are limited to 50 MHz. A reference RO used for comparison purposes was previously fabricated in the 28-nm FD SOI technology node. It was constructed with 44-stage inverters and a nominal oscillating frequency of 1 GHz.

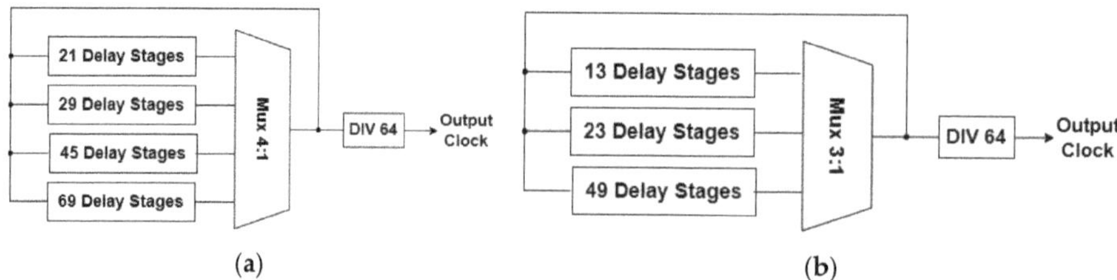

(a)   (b)

Figure 2. Schematic designs of different gate-based ROs: (**a**) Inverter-based RO schematic design; (**b**) NAND2 or NOR2-gate-based RO schematic design.

2.2. Flip-Flop Chains

The designed FF block included 14 different FF chains, which were comprised of a conventional master-slave transmission gate FF and 13 different radiation-hardened FF designs. The FFs were connected as shift register chains with 12,000 stages in each chain. The shift registers were clocked through an external clock signal and received input data via an IO pad. The reversed clock scheme was used to help avoid hold-time violations, as shown in Figure 3. While the power draw of each FF chain is slightly different due to the inherent design differences, together they offer an appropriate testing platform for determining changes in IC power usage as the dose rate increases, as well as changes in device functionality.

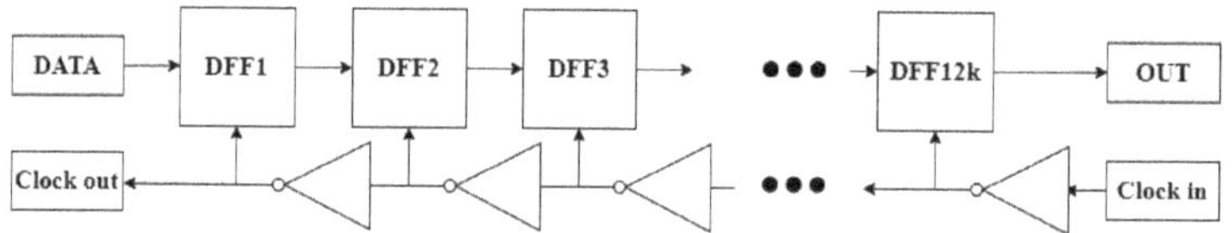

Figure 3. Flip-flop chain and the clock data flow for the shift register design.

2.3. SRAM Block

A forward body bias (FBB) transistor configuration was chosen for the SRAM array, as shown in Figure 4a. The FBB transistor configuration allows for higher drive and lower threshold voltages, whereas the conventional reverse body bias (RBB) configuration, shown in Figure 4b, can limit leakage currents by increasing the threshold voltage of the device. Traditionally, SOI technologies were prone to TID-induced leakage currents causing operational failures [17]. However, recent FD SOI technologies, such as 28-nm and 22-nm, have shown a significantly higher tolerance for TID exposures with limited increases in leakage currents, especially when the total absorbed dose is within 100 krad(Si) [18]. In this case, the usage of the RBB configuration to limit the operational performance of ICs was not as appealing as before. Instead, the usage of the FBB configuration to improve the performance of an IC design has become more valuable than the traditional RBB configuration. To investigate TID effects on an SRAM with FBB configuration, a 256 kbit SRAM block was designed with a memory compiler. The SRAM block featured a 15-bit address line, an 8-bit data line, as well as a read/write enable and clock input. It was a single-port SRAM configured as 32 K × 8 memory with 256-cells on a bit-line. It included additional features such as bit-line redundancy, a pipeline mode, and power-gating. Inside of the SRAM design, it included cell arrays and various peripheral circuits such as row/column decoders, self-timing generators, sense amplifiers, and buffers. As SRAMs are commonly used in circuit designs, it is important to evaluate their performance under TID effects.

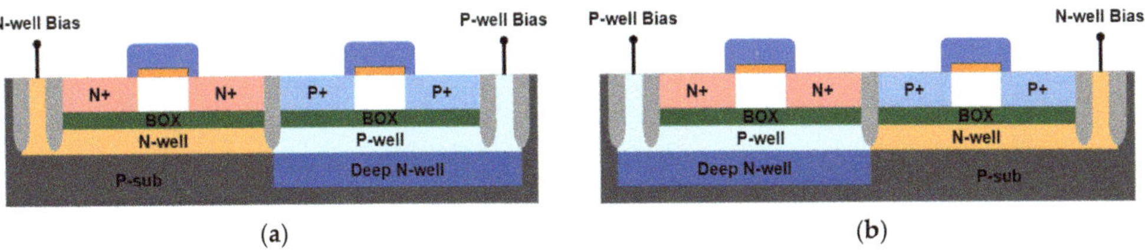

Figure 4. (a) Flipped well configuration (FBB); (b) regular well configuration (RBB).

3. Test Chip Design

A test chip consisting of the three types of circuits previously discussed was developed and fabricated in a commercial 22-nm FD SOI technology, as shown in Figure 5. All FF chains shared the same data input. The outputs of the SRAM block and FF chains were connected to the IO pads for external error detection, and the outputs of the RO circuits were connected to the IO pads for the frequency measurements. The nominal core logic supply voltage for this technology was 0.8 V, and the IO voltage was 1.8 V. Functional verification testing was carried out on the fabricated SRAM and FF chains with all 0s, 1s, and checker-board input patterns before irradiation.

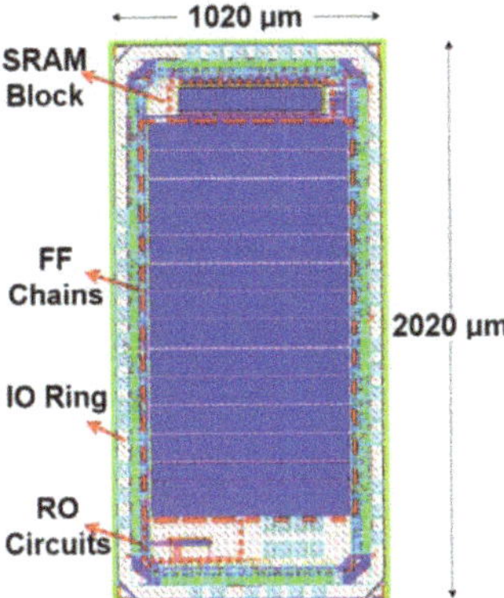

Figure 5. Test chip overall layout.

A testing system consisting of power supplies, an FPGA board, and a microcontroller used during testing. The test chip was soldered onto a custom-made daughter board and was connected to the FPGA via a DIMM connection. During testing, the FFs, SRAM, and ROs were powered at their nominal voltage of 0.8 V. The FF chains were clocked at 1 MHz with the 'all 0' data input pattern, and current readings were taken from the power supplies every minute. The SRAM block was also clocked at 1 MHz, and a checker-board data pattern was used. At the beginning of testing, all addresses in the SRAM were written with the test data. Then, the content of the SRAM was continually read out to check if the data were still appropriately stored. If there was a mismatch in data, the system would record the event and attempt to repair the address with the correct test data, and the process would continue. If the address data could not be repaired, then that address location was considered non-functional and recorded.

During testing, the ROs were also powered at their nominal voltage of 0.8 V. The output pins of the ROs were connected to counters inside the FPGA and monitored the number of oscillations of each RO within a 0.1 s period. By knowing the number of oscillations for a given period, the frequency of each RO was able to be determined. During testing, all collected data on the FPGA were transferred to a microcontroller via a serial connection, where the data were logged and recorded so that testing personnel could evaluate the experimental data in real-time.

4. TID Experimental Results and Discussions

The TID experiments were performed by using a Gammacell 220 Co-60 chamber (Figure 6) at the University of Saskatchewan, Saskatoon, SK, Canada. The Gammacell 220 chamber can provide an irradiation rate of 108.2 rad(Si) per minute. The total absorbed dose during the experiments was 1 Mrad.

Figure 6. Gammacell 220 Co-60 Irradiator.

The frequencies of the ROs were recorded during the TID testing and plotted in Figure 7. It should be noted that the observed experimental data are a linear trend, so a first-order polynomial curve fitting was implemented to understand this trend better. It is interesting to note that the ROs with the same gate type experienced the same decreasing rate in their output frequency during testing. This can simply be attributed to the same gate delay degradation during TID testing.

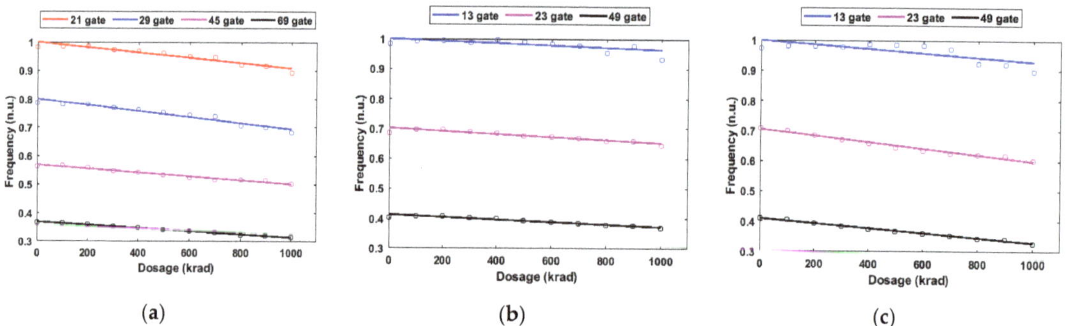

Figure 7. RO frequencies vs. total absorbed dose during Co-60 test: (**a**) Inverter-based ROs; (**b**) NAND2-based ROs; (**c**) NOR2-based ROs.

The relative change in frequency was calculated from the ROs designed with inverters, NAND2, and NOR2 gates and listed in Table 4. For the inverter-based RO, the relative decrease in frequency was the average of the drops in four different stage options; for the NAND2 and NOR2-based ROs, the average value of the drops in three different stage options was used to calculate the relative decrease in frequency. From the results, a trend emerged showing that ROs built using different gates were impacted by TID effects at differing rates.

Table 4. RO frequency differences from 0 krad(Si) of exposure to 1000 krad(Si) during Co-60 test.

RO Type	Relative Decrease in Frequency
Inverter	−12.3%
NAND2	−6.8%
NOR2	−14.1%

Table 4 shows that the frequencies of the NAND2 ROs were least affected by the TID, whereas the NOR2 ROs were most affected. It is noted that the NAND2 gates have two NMOS transistors in series and two PMOS in parallel, whereas the NOR2 gates have two PMOS transistors in series and two NMOS in parallel. The sizing of the NMOS and PMOS network of the NAND2 and NOR2 gates used in the ROs was designed such that the rising and falling time were roughly equal. In general, the positive charge trapped in the BOX under the transistors due to TID irradiation results in a negative shift of the threshold voltage, which leads to an increased driving current of the NMOS transistor and reduced driving current for the PMOS transistors in the gates. Findings in [2] show that the PMOS driving current is significantly reduced compared to the increase in the NMOS driving current in 28-nm FD SOI technology, which leads to the monotonic decrease in the frequency with the irradiation dose. The experimental results in this paper also show the same trends. The results of the NAND2 and NOR2 ROs further validate this since NOR2 has two PMOS transistors in series, which further degrades the delay of the gate, and hence reduces the frequency of the ROs.

TID testing with heavy ion irradiation was conducted at the Texas A&M University (TAMU) Cyclotron Institute. An irradiation rate of 60.9 rad(Si) per second was provided. Figure 8 shows the change of the frequencies as the increase in the heavy ions TID exposure for the inverter-based RO (45 stages), NAND2-based RO (23 stages), and NOR2-based RO (23 stages). The results still show that the NOR2 RO has the largest frequency degradation, while the NAND2 RO has the least, which was demonstrated from the previous analysis. In addition, it can be observed that the degradation of the RO frequency from the Co-60 test is much more than the heavy ion test. Despite the use of the High-K dielectric gates in advanced technologies, which can help to reduce the radiation-induced voltage shift in the gate insulator, the radiation-induced charge in SOI buried oxides and shallow trench isolation oxide (STI) can cause degradation or failures as well [19]. Electron-hole pairs are generated in the oxide layers when applying high-energy ionizing radiation, but most holes and electrons can immediately recombine. For high LET particles, such as heavy ions, they generate high-density charge pairs, making the initial recombination rate significantly large. The charge pair line density is relatively small for the low LET particles, such as Co-60, which reduces the initial recombination rate. Compared to the heavy ions, the Co-60 has a better ability and efficiency to create the trapped charges in oxide layers [9], as reported in [20,21], respectively.

Another inverter-based RO in 28-nm technology was used for comparison. There are 45 stages in the 28-nm RO, and 44 of them are also used as delay stages contributing to other designs. A multiplexer was used to switch between these two modes, as shown in Figure 9. When the select input is high, the design will work as an RO. The 28-nm RO was also based on the conventional inverter without any layout optimizing techniques, which is the same as the 22-nm design, making these two designs fully comparable. The frequency versus total dose for two inverter-based ROs (45 stages and 69 stages, respectively) in the 28-nm and 22-nm FD SOI technologies are plotted in Figure 10. These results show that during the Co-60 test at 850 krad(Si) of dosage, the 28-nm RO had a frequency decrease of over 30% from the initial, compared to the 14.5% decrease in 22-nm RO. During the heavy ion test, at 200 krad(Si) of dosage, the 28-nm RO had a frequency decrease of around 1.8% from the initial, compared to a less than 0.5% decrease in the 22-nm RO. These results show that the degradation in frequency for the 22-nm inverter-based RO has significantly improved. This indicates that the PMOS driving current is less affected by the TID effects for the 22-nm technology node, which could be due to manufacturing improvements. It is known that a thinner SiO_2 BOX layer can lead to a milder TID effect [18,22]. The thicknesses of the BOX and the SOI body of the 28-nm FD SOI technology are 25-nm and 7-nm [23], and those are 20-nm and 6-nm in the 22-nm FD SOI technology node, respectively [24]. The BOX of the 22-nm process is 25% thinner than that of the 28-nm process, so a less positive charge will be deposited in the BOX of the 22-nm process during irradiation. This will cause less interference with the threshold voltage of the 22-nm process, eventually leading to better resilience to TID effects.

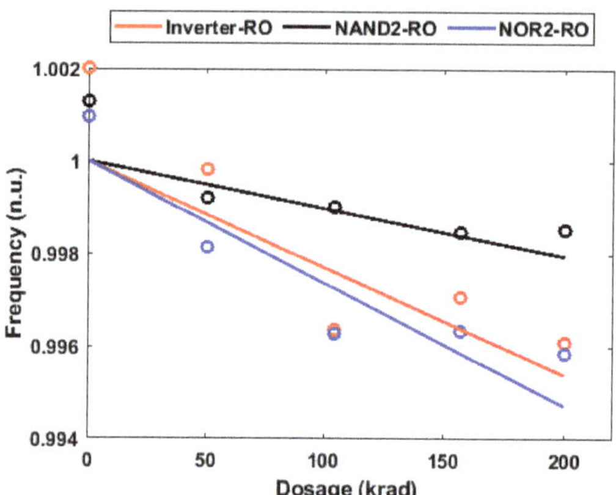

Figure 8. RO frequencies vs. total absorbed dose during heavy ion test.

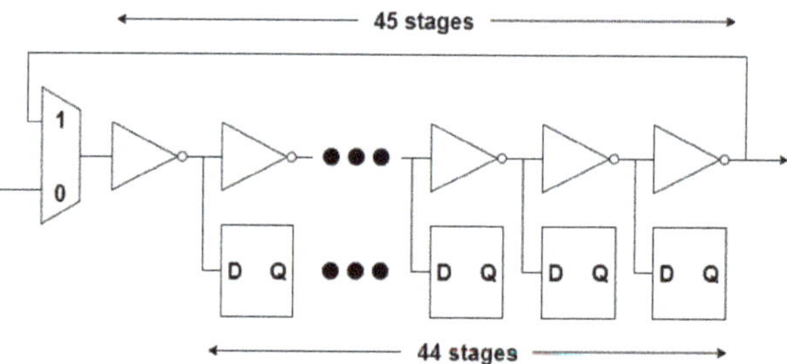

Figure 9. Schematic of the 28-nm RO design.

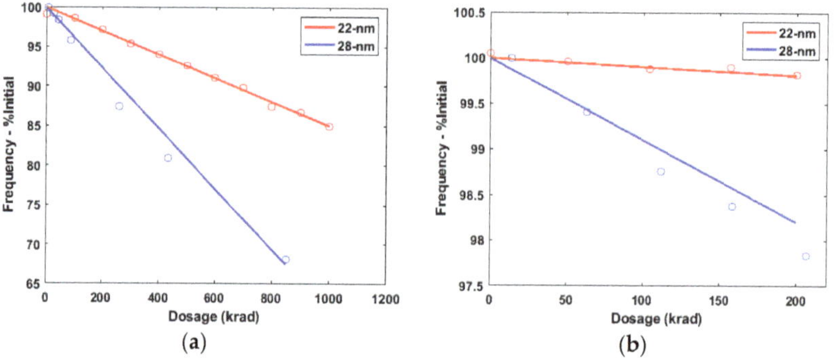

Figure 10. RO frequencies from 22-nm and 28-nm test for (**a**) Co-60 test; (**b**) heavy ion test.

Figure 11 shows the change in power supply current as the TID exposure increased for both 22-nm and 28-nm test circuits for Co-60 irradiation. Both of the plots are normalized. The current of 22-nm influenced by cumulative TID exposure is very similar to that of the 28-nm test. However, the current of the 22-nm test increased by approximately 18 times after 800 krad(Si) dosage compared to the initial starting current, while the 28-nm showed a 10 times increase. The increase in driving and leakage currents in the transistors can explain this increase in the current draw as the dose rate increases. These data show that TID effects are still prominent in the 22-nm FD SOI node. In addition, in the 22-nm technology node, the increase in the leakage current shows a linear trend from low doses to high doses.

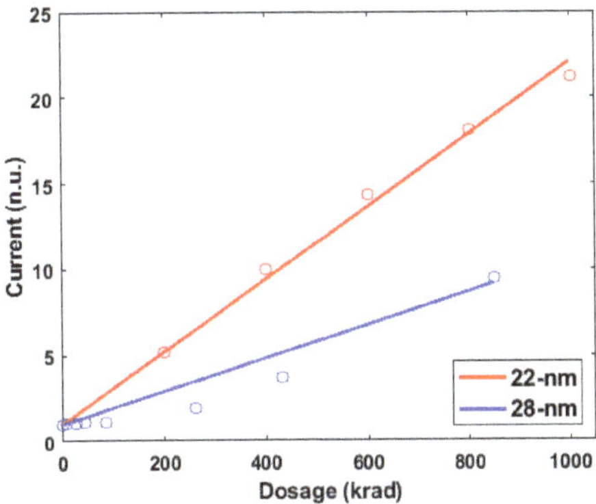

Figure 11. Current from 22-nm and 28-nm tests during Co-60 test.

The SRAM operated normally during testing until its functionality deteriorated at 325 krad(Si) of the total dosage. At this point, the number of recorded errors increased drastically, and every address showed a loss in functionality by 332 krad(Si) of the total dosage. A series of troubleshooting tests were performed to try and return functionality back to the SRAM block. These included resetting the test program, power cycling the SRAM, and clocking the SRAM at a slower rate of 100 kHz. None of these methods yielded any results, and the SRAM block was presumed dead. The FFs were still functional after the 1000 krad(Si) total absorbed dose, and no error was observed on any of the FF chains. In this case, it is believed that the SRAM storage cells themselves did not fail at 325 krad(Si), but instead, some part of the peripheral circuitry became damaged. For example, the internal operation of the SRAM block is controlled by a self-timing, asynchronous circuit, which could fail due to the increased delay, and this would explain why the entirety of the SRAM block failed at the same time instead of a more gradual loss in functionality. In this case, this timing issue caused by TID should be taken into account when designing the memory. For example, the self-timing circuit in the SRAM should be designed to be tolerant of the additional delay induced by the TID. Another method is to apply the back-gate voltage to the peripheral circuitry. When applying the back-gate voltage to circuits in the flipped well configuration, the drive current will be enhanced, and the delay will be reduced, which should effectively mitigate the TID effect in the peripheral circuitry.

5. Conclusions

Three types of RO circuits were designed and fabricated to evaluate the effects of frequency and leakage currents while exposed to gamma radiation. Results were compared with the previous 28-nm FD SOI node, and they show that the 22-nm FD SOI node had less frequency degradation compared to the 28-nm node. Among the three types of ROs, all showed degradations due to the TID, but the NOR ROs yielded a worse performance than that of the inverter ROs and NAND ROs due to the significantly reduced driving current in PMOS transistors during TID testing. The reduction rate of the RO frequency during the Co-60 was much higher than that of the HI test. The power supply current increased as the dosage increased similarly to that of the 28-nm node. The SRAM failed when the cumulated dose reached 325 krad(Si), but the FF chains were still functional through the test up to 1000 krad(Si). These results indicate that additional attention must be paid when designing complex circuits for radiation-hardened applications.

Author Contributions: Conceptualization, L.C., C.J.E. and Z.L.; methodology, Z.L.; software, Z.L.; validation, C.J.E. and Z.L.; formal analysis, Z.L.; investigation, C.J.E., Z.L., C.J., J.X., Z.Y. and S.S.; resources, L.C.; data curation, C.J.E. and Z.L.; writing—original draft preparation, Z.L.; writing—review and editing, L.C.; visualization, L.C.; supervision, L.C.; project administration, L.C.; funding acquisition, L.C. All authors have read and agreed to the published version of the manuscript.

Funding: This research was funded by NSERC and the Cisco University Research Program. This is also in part supported by the Beijing Microelectronics Technology Institute (TX-P21-01).

Institutional Review Board Statement: Not applicable.

Informed Consent Statement: Not applicable.

Data Availability Statement: The data presented in this study are available on request from the corresponding author. The data are not publicly available due to licensing agreements.

Acknowledgments: We appreciate the support from CMC Microsystems for accessing the technology PDK and the chip space.

Conflicts of Interest: The authors declare no conflict of interest.

References

1. Zhang, H. Effects of total-ionizing-dose irradiation on single-event response for flip-flop designs at a 14-/16-nm bulk FinFET technology node. *IEEE Trans. Nucl. Sci.* **2018**, *65*, 1928–1934. [CrossRef]
2. Seo, D.; Trang, L.D.; Han, J.-W.; Kim, J.; Lee, S.; Chang, I.-J. Total ionizing dose effect on ring oscillator frequency in 28-nm FD-SOI technology. *IEEE Electron. Device Lett.* **2018**, *39*, 1728–1731. [CrossRef]
3. Zheng, Q. Total ionizing dose responses of forward body bias ultra-thin body and buried oxide FD-SOI transistors. *IEEE Trans. Nucl. Sci.* **2019**, *66*, 702–709. [CrossRef]
4. King, M.P. Analysis of TID process, geometry, and bias condition dependence in 14-nm FinFETs and implications for RF and SRAM performance. *IEEE Trans. Nucl. Sci.* **2017**, *64*, 285–292. [CrossRef]
5. Gaillardin, M. Total ionizing dose response of multiple-gate nanowire field effect transistors. *IEEE Trans. Nucl. Sci.* **2017**, *64*, 2061–2068. [CrossRef]
6. Riffaud, J. Investigations on the geometry effects and bias configuration on the TID response of nMOS SOI tri-gate nanowire field effect transistors. *IEEE Trans. Nucl. Sci.* **2018**, *65*, 39–45. [CrossRef]
7. Hughes, H. Total ionizing dose radiation effects on 14 nm FinFET and SOI UTBB technologies. In Proceedings of the 2015 IEEE Radiation Effects Data Workshop (REDW), Boston, MA, USA, 13–17 July 2015.
8. Paillet, P. Total ionizing dose effects on deca-nanometer fully depleted SOI devices. *IEEE Trans. Nucl. Sci.* **2015**, *52*, 2345–2352. [CrossRef]
9. Liu, R. Single event transient and TID study in 28 nm UTBB FDSOI technology. *IEEE Trans. Nucl. Sci.* **2017**, *64*, 113–118. [CrossRef]
10. Gaillardin, M. Impact of SOI substrate on the radiation response of ultrathin transistors down to the 20 nm node. *IEEE Trans. Nucl. Sci.* **2013**, *60*, 2583–2589. [CrossRef]
11. Javanainen, A. Heavy-ion induced charge yield in MOSFETs. *IEEE Trans. Nucl. Sci.* **2009**, *56*, 3367–3371. [CrossRef]
12. Trang, L.D. Comparing variation-tolerance and SEU/TID-resilience of three SRAM cells in 28nm FD-SOI technology: 6T, Quatro, and we-Quatro. In Proceedings of the 2020 IEEE International Reliability Physics Symposium (IRPS), Grapevine, TX, USA, 28 April–30 May 2020.
13. Feeley, A.; Xiong, Y.; Guruswamy, N.; Bhuva, B.L. Effect of frequency on total ionizing dose response of ring oscillator circuits at the 7-nm bulk FinFET node. *IEEE Trans. Nucl. Sci.* **2022**, *69*, 327–332. [CrossRef]
14. Fujishima, M.; Asada, K. Proposal of standard characterization method for dynamic circuit performance. In Proceedings of the 1993 International Conference on Microelectronic Test Structures, Sitges, Spain, 22–25 March 1993.
15. Bhushan, M.; Gattiker, A.; Ketchen, M.B.; Das, K.K. Ring oscillators for CMOS process tuning and variability control. *IEEE Trans. Semicond. Manuf.* **2006**, *19*, 10–18. [CrossRef]
16. Zafarkhah, E.; Maymandi-Nejad, M.; Zare, M. Improved accuracy equation for propagation delay of a CMOS inverter in a single ended ring oscillator. *Int. J. Electron. Commun.* **2017**, *71*, 110–117. [CrossRef]
17. Rezzak, N. Total-ionizing-dose radiation response of 32 nm partially and 45 nm fully-depleted SOI devices. In Proceedings of the 2012 IEEE International SOI Conference (SOI), Napa, CA, USA, 1–4 October 2012.

18. Yan, G. Simulation of Total Ionizing Dose (TID) effects mitigation technique for 22 nm Fully-Depleted Silicon-on-Insulator (FDSOI) transistor. *IEEE Access* **2020**, *8*, 154898–154905. [CrossRef]
19. Schwank, J.R. Radiation effects in MOS oxides. *IEEE Trans. Nucl. Sci.* **2018**, *55*, 1833–1853. [CrossRef]
20. Shaneyfelt, M.R.; Fleetwood, D.M.; Schwank, J.R.; Hughes, K.L. Charge yield for cobalt-60 and 10-keV X-ray irradiations of MOS devices. *IEEE Trans. Nucl. Sci.* **1991**, *38*, 1187–1194. [CrossRef]
21. Dudek, P.; Szczepanski, S.; Hatfield, J.V. A high-resolution CMOS time-to-digital converter utilizing a Vernier delay line. *IEEE J. Solid-State Circuits* **2000**, *35*, 240–247. [CrossRef]
22. Gaillardin, M. Radiation effects in advanced SOI devices: New insights into total ionizing dose and single-event effects. In Proceedings of the 2013 IEEE SOI-3D-Subthreshold Microelectronics Technology Unified Conference (S3S), Monterey, CA, USA, 7–10 October 2013.
23. Zhang, K.; Kanda, S.; Yamaguchi, J.; Furuta, J.; Kobayashi, K. Analysis of BOX layer thickness on SERs of 65 and 28nm FD-SOI processes by a monte-Carlo based simulation tool. In Proceedings of the 2015 15th European Conference on Radiation and Its Effects on Components and Systems (RADECS), Moscow, Russia, 14–18 September 2015; pp. 1–5. [CrossRef]
24. Yuan, J.; Zhao, Y.; Wang, L.; Li, T.; Sui, C. 22nm FDSOI SRAM single event upset simulation analysis. *J. Phys. Conf. Ser.* **2021**, *1920*, 12069. [CrossRef]

Article

Quantitative Research on Generalized Linear Modeling of SEU and Test Programs Based on Small Sample Data

Fei Chu [1,2], Hongzhuan Chen [1], Chunqing Yu [2,*], Lihua You [2], Liang Wang [2] and Yun Liu [3]

1   Department of Economics and Management, Nanjing University of Aeronautics and Astronautics, Nanjing 211106, China; sdcf_2000@163.com (F.C.); 13813922476@163.com (H.C.)
2   Beijing Microelectronics Technology Institute, Beijing 100076, China; youlihua08@163.com (L.Y.); wangliang150200@163.com (L.W.)
3   College of Engineering and Computer Science, Australian National Unversity, Canberra 2600, Australia; caiyimao101@163.com
*   Correspondence: yuchunqing2006@163.com

Abstract: Complex integrated circuits (ICs) have complex functions and various working modes, which have many factors affecting the performance of a single event effect. The single event effect performance of complex ICs is highly program-dependent and the single event sensitivity of a typical operating mode is generally used to represent the single event performance of the circuits. Traditional evaluation methods fail to consider the cross effects of multiple factors and the comprehensive effects of each factor on the single event soft error cross section. In order to solve this problem, a new quantitative study method of single event error cross section based on a generalized linear model for different test programs is proposed. The laser test data is divided into two groups: a training set and a validation set. The former is used for model construction and parameter estimation based on five methods, such as the generalized linear model and Ensemble, while the latter is used for quantitative evaluation and validation of a single event soft error cross section of the model. In terms of percentage error, the minimum mean estimation error on the validation set is 13.93%. Therefore, it has a high accuracy to evaluate the single event soft error cross section of circuits under different testing programs based on the generalized linear model, which provides a new idea for the evaluation of a single event effect on complex ICs.

Keywords: processor; laser test; generalized linear model; ensemble method; confidence interval

Citation: Chu, F.; Chen, H.; Yu, C.; You, L.; Wang, L.; Liu, Y. Quantitative Research on Generalized Linear Modeling of SEU and Test Programs Based on Small Sample Data. *Electronics* **2022**, *11*, 2242. https://doi.org/10.3390/electronics11142242

Academic Editor: Paul Leroux

Received: 29 May 2022
Accepted: 17 July 2022
Published: 18 July 2022

Publisher's Note: MDPI stays neutral with regard to jurisdictional claims in published maps and institutional affiliations.

1. Introduction

With the rapid development of aerospace technology and deeper exploration in space, the requirements for performance of spacecraft are also increasing. Correspondingly, the reliability and the radiation-hardened performance of complex integrated circuits (ICs) are facing higher requirements [1]. Nowadays, with the reduction in the process size of semiconductor devices, the single event effect (SEE), is seriously affecting the safety of space missions, which is becoming more and more significant [2]. Therefore, it is necessary to evaluate the SEE sensitivity of complex ICs before they are applied to space missions [3]. At present, the SEE evaluation methods of ICs recognized by the industry mainly include a heavy ion test, a proton test, and other radiation tests. The SEE sensitivity of complex ICs has a strong program dependence on which different users have different concerns and applications, so the test programs cannot be traversed. In addition, due to the limited time and high cost of the heavy ion accelerator, it is not suitable for all ICs to carry out radiation tests. Therefore, several simulation methods and fault injection methods have been used to study the SEE sensitivity of complex ICs.

VSenek et al. [4] proposed a Single Event Upset (SEU) simulation prediction method based on the duty cycle to predict the SEU cross section of processors under different test programs. A simple error rate prediction model was preliminarily established. However, the method of analyzing the duty factor was difficult to be applied to complex applications, such as programs with conditional branches. Emmanuel et al. [5] carried out the SEU simulation on processor under different programs based on fault injection, and applied a new fault model of multi-fault injection to dual-fault injection. The model was designed to represent a possible non-concurrent radiation-induced soft error, which was only useful for specific processors in this paper. Gao jie, li qiang et al. [6] studied the relationship between the dynamic and the static SEU rates of satellite microprocessors by using the concept of program duty ratio and fault injection technology, which verified only by fault injection but not by radiation experiments. Zhao Yuanfu et al. [7] proposed a method to predict the SEE of complex ICs, which required a detailed analysis of different test programs and a large amount of work. The prediction method has been verified by radiation experiments, which has a good guiding significance.

The simulation method needs to establish different models for different circuits, which is complicated and time-consuming, and the simulation results have relatively large errors compared with the real test results. The method of fault injection has some problems of precision, accuracy, and speed for the modeling. None of the above methods have taken the cross influence of multiple factors and the comprehensive influence and accurate quantification of SEE soft error cross section into account.

In view of the above shortcomings, laser SEE test data as small sample training set has been used for modeling based on the generalized linear model. Quantitative evaluation on the SEE cross section of the circuit under different test programs has been conducted and the evaluation errors of training set and validation set under different methods are verified, and they have been compared. It provides a new idea for the evaluation of SEE soft error cross section in complex ICs. The SEU cross section of devices under different test programs can be predicted by the new method without carrying out radiation tests on all test modes. It can effectively solve the problem of evaluating the radiation performance of complex ICs under different test modes and obtain high accuracy.

2. Circuit Descriptions and Radiation Experiments

2.1. Circuit Description

The research object is a 32-bit radiation-hardened microprocessor, which has all the typical characteristics of complex ICs, such as large scale, high frequency, multiple modules and complex functions. The system-level error detection and correction are adopted by the radiation-hardened microprocessor. The circuit consists of an integer processing unit (IU), a floating point processing unit (FPU), CACHE, register (REGFILE), a debugging support unit (DSU), a serial port (UART), a storage/interrupt controller, a watchdog, a timer, and other units, which realize data interaction through AMBA bus. The functional block diagram of a microprocessor is shown in Figure 1.

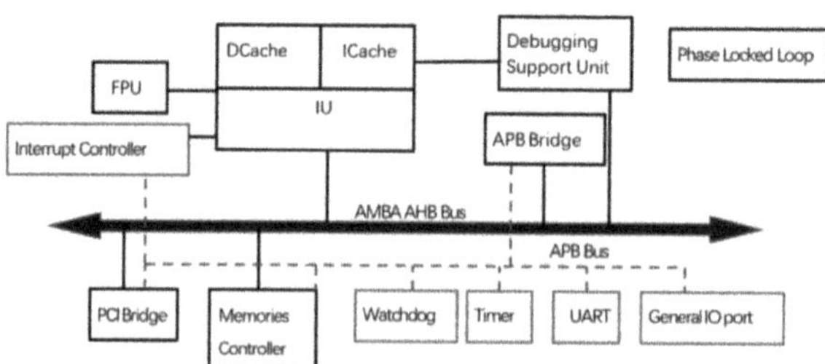

Figure 1. Functional block diagram of the microprocessor.

2.2. Experiment Setting

In this paper, a set of function test programs is developed to simulate the typical function state of a user, which makes the CPU instruction coverage reach 100%, and it also covers all the logic units of the circuit. Taking P1–P4 test programs as the training test, they perform a single-precision integer operation and a double-precision floating point operation in the CACHE open and closed states, respectively, which cover eight standard test functions. P1 is a single-precision integer operation in CACHE ON mode, P2 is a single-precision integer operation in CACHE OFF mode. P3 is the double-precision floating-point operation in CACHE ON mode, and P4 is the double-precision floating-point operation in CACHE OFF mode. P5–P8 test programs are designed as validation test programs. In order to better verify the effectiveness of the model, this validation program has randomness, and it does not require instruction coverage and logical unit coverage. The processor will continuously access Refile data when executing different test programs. The statistics of register usage of the target circuit under eight different test programs are shown in Table 1.

Table 1. Statistics of register usage under eight test programs.

Programs	Register Reads (Times)	Program Execution Cycle T1 (s)	Average Register Access Time T2 (us)
P1	~29,843	0.063	2.11
P2	~29,843	0.714	23.93
P3	~327,660	0.072	0.22
P4	~327,660	0.857	2.62
P5	~74,468	0.330	17.48
P6	~71,015	0.837	2.45
P7	~301,430	0.524	21.20
P8	~193,327	0.247	14.62

2.3. Radiation Experiments

The single event upset soft error cross sections (SEU cross sections) of the circuit under different test programs are obtained by pulse laser test [8]. During the test, the working voltage of the target circuit is set as the lowest level, where IO voltage is 2.97V and core voltage is 1.62 V. The backside irradiation laser test is carried out using PL2210A-P17 pulsed laser with 100 Hz frequency. The laser SEE test site is shown in Figure 2. The laser test data of the training set is shown in Table 2, and the laser test data of the validation set is shown in Table 3 in which the effective laser energy (i.e., laser energy focused in the active region E_{eff}) has been equivalent to the LET value of heavy ion [9].

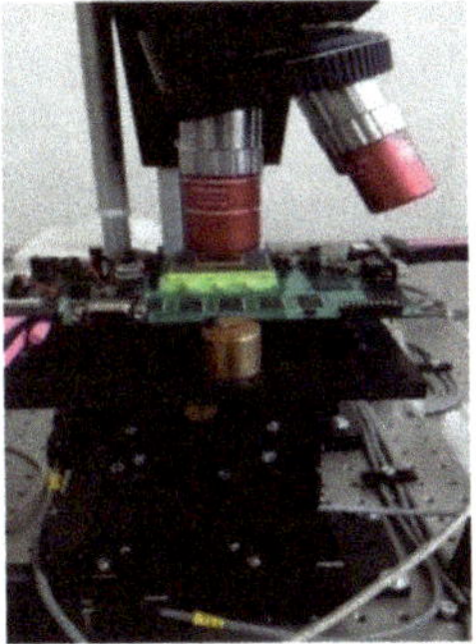

Figure 2. Laser SEE test site.

Table 2. Laser SEE soft error cross section of Training set (10^{-7} errors/cm^2).

Test Programs	Equivalent LET	324pJ = 28.7 (MeV.cm^2/mg)	432pJ = 37.5 (MeV.cm^2/mg)	621pJ = 53 (MeV.cm^2/mg)	950pJ = 80 (MeV.cm^2/mg)
P1		11	31.8	51.7	114.3
P2		6.3	22	33	69.9
P3		16.9	46.5	72	147.8
P4		9.3	24.4	35	76.8

Table 3. Laser SEE soft error cross section of validation set (10^{-7} errors/cm^2).

Test Programs	Equivalent LET 135pJ = 13.1 (MeV.cm^2/mg)	241pJ = 21.8 (MeV.cm^2/mg)	241pJ = 21.8 (MeV.cm^2/mg)	889pJ = 75 (MeV.cm^2/mg)
P5	18.1	23.5	33.6	112.2
P6	21.1	16.5	26.0	59.6
P7	21.9	14.7	39.5	131.7
P8	4.8	11.7	15.2	70.8

The conversion relation between the initial laser energy E_0 and the LET value of the heavy ion is shown in Formula (1):

$$\begin{cases} E_{eff} = f(1 - R)e^{-\alpha h}(1 + R')E_0 \\ \quad LET = 0.082E_{eff} + 2.07 \end{cases} \tag{1}$$

where f is the effect factor of the spot, R is the reflectance of the device surface, R' is the metal layer reflectance of the device, α is the silicon substrate absorption coefficient of the device and its measured value, and h is the Planck constant.

Four validation programs for P5–P8 were designed, and laser tests were carried out under different laser energies, respectively. The obtained laser test data of P5–P8 are shown in Table 3.

3. Modeling and Parameter Estimation

Compared with the data obtained by software simulation, the laser test data is closer to the radiation sensitivity of the circuit in the actual radiation environment, so the model established based on laser test data have a higher accuracy. There are 16 observations of laser test data, which are typical small sample data, and the use of second-order and above polynomial or tree models may lead to overfitting [10–12]. The generalized linear model is linear with respect to the unknown parameters, but nonlinear with respect to the known variables. The nonlinear functional relational quantization model between the independent variable and the dependent variable can be established based on the linear parameters and multiple bases. The corresponding model has a good fitting effect and prediction accuracy. Therefore, the pulsed laser SEE test data of P1–P4 as the training set are used to build the model based on the generalized linear model. Four methods of generalized least squares method [10,11], the weighted least squares estimation method [13], the median regression method [12], and the least trimmed squares method [12] are used to obtain the estimation parameter of the generalized linear model. The four methods mentioned above are combined with optimal weights, namely the Ensemble method as the fifth method. Then, a laser test is performed on the target circuit under the P5–P8 validation programs, and the obtained laser test data is used as the validation set to verify the generalized linear model. The flow chart of the method to quantitatively evaluate the SEE soft error cross section of complex ICs based on the generalized linear model is shown in Figure 3.

Firstly, the P1–P4 test program for the training set is written, and 16 groups of small sample test data are obtained by laser test. The test data is used as the training set for the generalized linear model. Then, GLS, WLS, MR, LTS, and Ensemble methods are established and parameters are optimized on the target functions. The P5–P8 test programs as the validation set are written and then the generalized linear model is used to predict the SEE soft error cross section. Meanwhile, the soft error cross section under radiation is obtained by laser test. By evaluating the prediction error and the confidence interval on the test data, the precision of the quantitative prediction model is obtained.

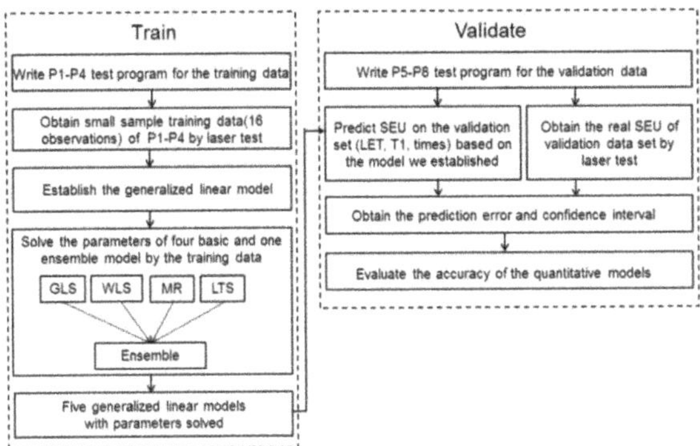

Figure 3. The flow chart of the method.

3.1. Model Building

The generalized linear model is established using Formula (2).

$$g(SEU) = \beta_0 + \sum_{i=1}^{p} \beta_i f_i(LET, times, T1, T2) + \varepsilon \tag{2}$$

Formula (2) represents the quantitative relationship between SEU and its impact factor, such as $LET, times, T1, T2$. Since the value of the SEU soft error cross section must be positive, and the right side of the equation takes the value of the entire real number domain, we take the logarithm of the SEU and put it into the model, that is, taken as the link function of Poisson regression. The model is linear with respect to unknown parameters, but nonlinear with respect to several known independent variables. Here, the function $f_i(LET, times, T1, T2)$ can be made of any nonlinear function about independent variables according to your observations and assumptions, such as neural networks, GBDT (gradient boosting decision tree), spline functions, or polynomials about independent variables, etc. [10,11]. In this paper, $f_i(LET, times, T1, T2)$ is taken as each independent variable itself to minimize the number of parameters to prevent overfitting and to achieve better predictions. The simulation software used for modeling and parameter estimation in this paper is R-3.5.3.

At this point, the above model can be simplified to Formula (3)

$$Y = X\beta + \varepsilon \tag{3}$$

Among them, $Y = \log(SEU) = (y_1, y_2, \ldots, y_n)^T$, n is the number of experiments, which is 16 in this paper, and SEU is a column vector of length n. Each column of X is the number of registers read times (times), program execution cycle (T1), average register access time (T2), LET and an intercept term (I) that is all 1, each row of X is an observation data. The dimension of X matrix is $n \times 5$. β is the unknown parameter column vector of length 5 to be solved, and ε is the measurement error column vector of length n.

3.2. Variable Selection

There are multiple criteria such as C_p, AIC, and BIC for variable selection of the model [11]. In this paper, AIC criterion is chosen for variable selection, which is the sum of the negative average log-likelihood and the penalty term considering the number of parameters. The lower the value, the better the prediction effect of the model. The result of variable selection using the AIC criterion is shown in Figure 4.

```
Start:  AIC=-30.36
I(log(SEU)) ~ LET + times + T1 + T2 + I + 0

         Df Sum of Sq    RSS     AIC
- T2      1     0.007   1.291 -32.28
<none>                  1.284 -30.36
- times   1     0.220   1.504 -29.83
- T1      1     0.719   2.003 -25.24
- I       1     3.649   4.933 -10.83
- LET     1     9.403  10.687   1.54

Step:  AIC=-32.28
I(log(SEU)) ~ LET + times + T1 + I - 1

         Df Sum of Sq    RSS     AIC
<none>                  1.291 -32.28
- times   1     0.393   1.683 -30.03
- T1      1     1.286   2.577 -23.22
- I       1     4.675   5.966  -9.79
- LET     1     9.403  10.694  -0.45

Call:
lm(formula = I(log(SEU)) ~ LET + times + T1 + I - 1, data = d_tr)

Coefficients:
        LET        times           T1            I
 0.03934239   0.00000106  -0.78625060   1.71685441
```

Figure 4. Variable selection using AIC criteria.

As can be seen from Figure 3, the final result of variable selection retains the number of register reads (times), program execution cycle (T1), LET and the intercept term (I) that is all 1, and it excludes the average register access time (T2).

3.3. Parameter Estimation

For different assumptions of measurement error or different loss functions, different parameter estimates $\hat{\beta}$ and predicted values of SEU cross section $S\hat{E}U$ can be obtained. In order to further analyze the experimental data and to obtain a more accurate and robust model, the following four methods are firstly used to carry out model building, in which the first two are based on the Gaussian distribution assumption, and the latter two are robust parameter estimation methods, both of them are used to solve Formula (3), and $\hat{\beta}$ is the estimated value of β. Methods for parameter estimation are shown in Table 4.

Table 4. Methods for parameter estimation.

Methods for Parameter Estimation	Formulas	Parameter Description
GLS, generalized least squares [10,11]	$\hat{\beta}_{GLS} = \underset{\beta}{argmin} \sum_{i=1}^{n} (Y_i - X_i^T \beta)^2 = (X^T X)^{-1} X^T Y$	X_i is the column vector formed for the i row of the matrix X.
WLS, weighted least squares [13]	$\hat{\beta}_{WLS} = \underset{\beta}{argmin} \sum_{i=1}^{n} w_i(Y_i - X_i^T \beta)^2 = (X^T W X)^{-1} X^T W Y$ $W = diag\{w_1, w_2, \ldots, w_n\}, w_i = \frac{1}{SEU_i}$	X_i is the column vector formed for the i row of the matrix X.
MR, median regression [12]	$\hat{\beta}_{MR} = \underset{\beta}{argmin} \sum_{i=1}^{n} |Y_i - X_i^T \beta|$	To calculate the $\hat{\beta}_{MR}$, slack variables can be introduced and the simplex method can be used [13].
LTS [12], least trimmed squares	$\hat{\beta}_{LTS} = \underset{\beta}{argmin} \sum_{i=1}^{h} |\varepsilon|_{(i)}^2$ $h = \left[\frac{n+p+1}{2}\right]$ $|\varepsilon|_{(1)} \leq |\varepsilon|_{(2)} \leq \cdots \leq |\varepsilon|_{(n)}$	p is the number of columns of X, $[m]$ represents the largest integer not greater than m. $(|\varepsilon|_{(1)}, |\varepsilon|_{(2)}, \ldots, |\varepsilon|_{(n)})^T$ is the vector sorted by absolute value from smallest to largest for each element in $\varepsilon = (\varepsilon_1, \varepsilon_2, \ldots, \varepsilon_n)^T$.

The estimation parameter for the training set obtained by the above four methods are shown in Table 5. It can be seen that the influence coefficients of LET and times obtained by different methods are positive, while T1 is on the contrary, and the values of the four methods are relatively close, that is, they increase with the growing of LET and times and decrease with the increase of T1.

Table 5. Estimation parameters for training set under different methods.

Column Names of X	GLS	WLS	MR	LTS
(Intercept)	1.71685	1.52232	2.34029	2.64312
LET	0.03934	0.04237	0.03053	0.02697
times	1.0578×10^{-6}	1.18×10^{-6}	8.26×10^{-7}	7.48×10^{-6}
T1	-0.78625	-0.79952	-0.78671	-0.83454

3.4. Model Optimization

Considering that each of the four methods in Section 3.3 may have advantages and disadvantages, the Ensemble method [10,11], which performs optimized linear weighting on the predicted values of these four methods by minimizing the combined variance, is adopted to reduce the prediction variance, shrink the confidence interval, and make predictions more robust. It can be seen from Formula (4) that after Ensemble, the variance of the forecast value, which is the weighted average of the evaluation from the four methods mentioned above, is reduced to minimize and thus becomes more reliable.

$$\min_{p} p^T \Sigma p$$
$$s.t. \ \sum_{i=1}^{4} p_i = 1 \tag{4}$$
$$p_i \geq 0$$

Σ is the covariance matrix of the evaluation errors of the four methods.

The covariance matrix of these four methods in Section 3.3 calculated based on the training dataset is shown in Table 6, where the diagonal is the variance of the evaluation errors of the four methods. It can be seen that the evaluation variances of the two robust parameter estimation methods, MR and LTS, are all less than 50, which means the standard deviation of the evaluation errors of the SEU cross section prediction value is less than $\sqrt{50} = 7.07$. The covariance of the WLS method and the LTS method is the smallest, which is -2.9, thus the correlation coefficient of these two is $\rho_{WLS,LTS} = \frac{-2.9}{\sqrt{230.93*48.53}} = -0.03$. The negative correlation of these two is used by the Ensemble method to further reduce the evaluation error variance of SEU.

Table 6. Covariance matrix of the evaluation errors of the four methods.

	GLS	WLS	MR	LTS
GLS	122.15	165.03	33.73	10.30
WLS	165.03	230.93	31.04	-2.90
MR	33.73	31.04	38.39	38.97
LTS	10.30	-2.90	38.97	48.53

The inner point method of constraint optimization algorithm [12,13] is used to calculate the optimal weights $p = \begin{Bmatrix} 0.00118461 \\ 0.09161878 \\ 0.68351429 \\ 0.22368231 \end{Bmatrix}$. It can be seen that MR method has the largest weight, while GLS method has almost zero weight.

3.5. Evaluation Error

In order to verify the evaluation performance of the model on the test data, the above five evaluation methods are used in the training set and the validation set to verify the evaluation effect of the model, respectively. For the evaluation of error, root mean square error (RMSE) [10,11], average error percentage, and other various evaluation indicators are selected. The smaller the value, the smaller the evaluation error and the higher the evaluation accuracy. The evaluation errors of the training set and the validation set under different methods are shown in Table 7.

Table 7. Evaluation errors of training set and validation set under different methods.

Method	Evaluation Errors of Training Set under Different Methods					Evaluation Errors of Validation Set under Different Methods				
	GLS	WLS	MR	LTS	Ensemble	GLS	WLS	MR	LTS	Ensemble
Root mean square error	10.74	14.87	6.22	8.03	6.40	8.92	11.11	6.44	7.57	6.41
Mean absolute error	9.11	11.31	4.38	5.05	4.76	7.58	9.34	5.03	6.03	4.83
Mean absolute error in percent %	22.38	30.98	12.96	16.73	13.33	19.39	24.15	14.0	16.4	13.93

Taking RMSE [10,11], the most commonly used evaluation method, as an example, Ensemble and the MR method have relatively low RMSE in the training set and the validation set, while the other three methods have relatively large evaluation errors in validation set. The MR method, as one of the robust parameter estimation methods, shows a lower evaluation error in the training set and the validation set compared with the GLS and the WLS methods based on Gaussian distribution, indicating that the error of SEU data collected in the experiment may be non-Gaussian distribution, or Gaussian characteristics are not obvious due to small samples. It also verified the reliability of the MR and the Ensemble methods in data modeling, that is, the quantitative evaluation model established by us is effective, and it has a certain accuracy.

4. Confidence Interval Analysis

In the above section, the evaluation errors are compared. Considering the quantity of data is small, the performance of five methods can be affected by accidental data, therefore in this section further analysis on the confidence interval is carried out. The shorter confidence interval means a smaller evaluation error variance and a higher reliability. So, in this section, further comparison of confidence interval with five methods is performed.

In this paper, the bootstrap method based on statistical sampling is used to calculate the confidence interval of SEU cross section. Multiple SEU cross sections can be obtained by repeated sampling with replacements. The bootstrap times in this paper is set as B = 300.

Figures 5–8 show the evaluation value and a 95% confidence interval of the five methods on the training set and the validation set, respectively. The evaluation error is the difference between the evaluation value and the soft error cross section value observed in the laser experiment. Gray bars representing known experimental values, points and line segments of five colors are the respective evaluation values and confidence intervals of the five methods. The narrower the confidence interval is, the higher the evaluation accuracy is. Figures 7 and 8 show the probability density function of the evaluation error of the five methods on the training set and validation set, which is calculated using kernel method (gaussian kernel is used, bandwidth is determined based on cross-validation). It can be seen that the value of the error corresponding to the highest density of Ensemble and MR methods on the validation set is concentrated near zero, but the former is more concentrated at zero than the latter. The error distribution of the other three methods is relatively flat, and the error corresponding to the highest probability density is far from zero. As can be seen from Figures 5–8 and Table 7, the Ensemble method has the best evaluation accuracy RMSE and the narrowest confidence interval in the validation set, which can be used to evaluate the SEE soft error cross section based on different test programs and laser energy.

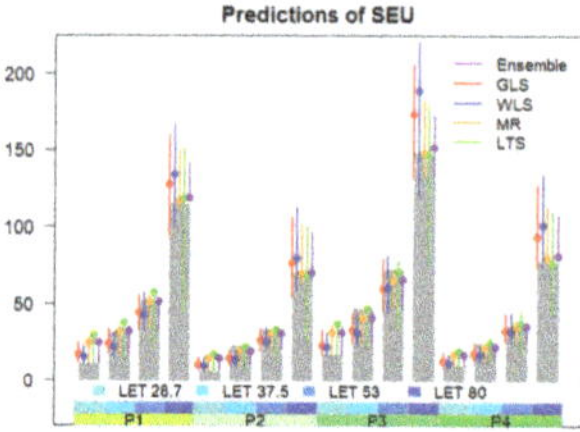

Figure 5. Evaluations, 95% confidence intervals and real values of the five methods on the training set.

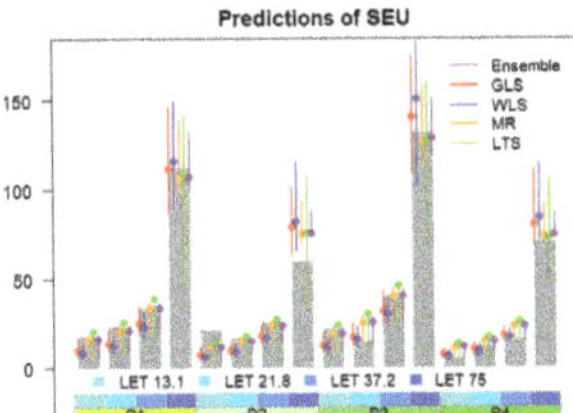

Figure 6. Evaluations, 95% confidence intervals and real values of the five methods on the validation set.

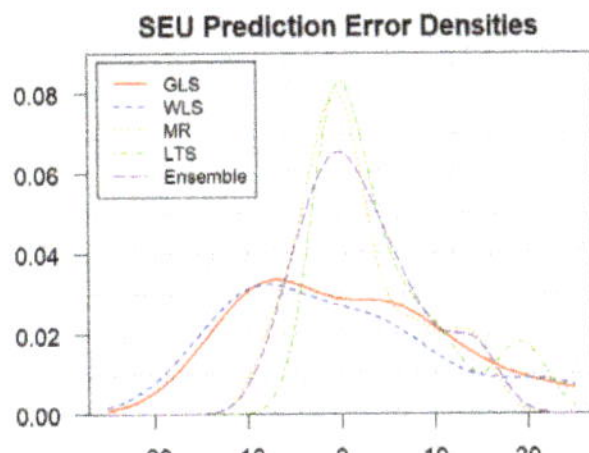

Figure 7. Density function of evaluation errors of five methods on training set.

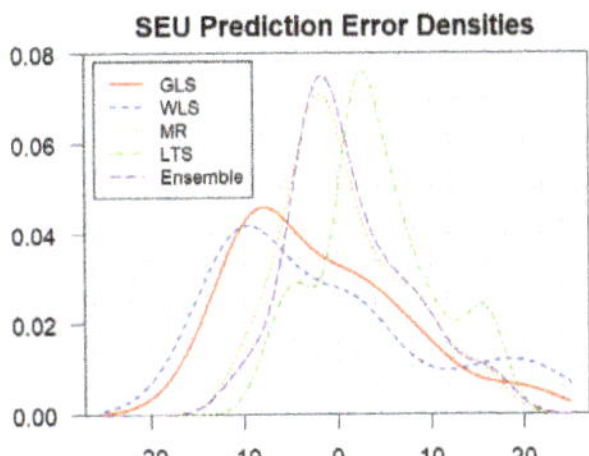

Figure 8. Density function of evaluation errors of five methods on validation set.

5. Conclusions

A new method based on the generalized linear model for quantitative evaluation of a SEE soft error cross section under different test programs is presented. Different test programs are designed, and the data sets of a laser test under several test programs and register calls of different test programs are divided into training and validation groups. The training set is used for modeling and parameter estimation of the five methods, and the validation set is used to evaluate the model accuracy. The evaluation value, evaluation error, 95% confidence interval, and the probability density function of the five methods on the training set and the validation set are computed. The results show that the quantitative evaluation method of complex ICs based on generalized linear models can achieve a high accuracy of 13.93%. Various factors of the SEE sensitivity of comprehensive effect and quantitative evaluation are considered in the evaluation method. Quantitative evaluation is suitable for small sample experiment data under different test programs, which is an applicable innovation.

Author Contributions: Conceptualization, F.C. and H.C.; methodology, F.C.; software, C.Y. and L.Y.; validation, F.C., L.W. and Y.L.; formal analysis, F.C.; investigation, F.C. and C.Y.; resources, C.Y.; data curation, L.Y.; writing—original draft preparation, F.C. and Y.L.; writing—review and editing, H.C. and L.W.; visualization, L.Y.; supervision, C.Y.; project administration, F.C. and H.C.; funding acquisition, F.C. All authors have read and agreed to the published version of the manuscript.

Funding: This research received no external funding.

Conflicts of Interest: The authors declare no conflict of interest.

References

1. Chen, R.; Zhang, F.; Chen, W.; Ding, L.; Guo, X.; Shen, C.; Luo, Y.; Zhao, W.; Zheng, L.; Guo, H.; et al. Single-Event Multiple Transients in Conventional and Guard-Ring Hardened Inverter Chains Under Pulsed Laser and Heavy-Ion Irradiation. *IEEE Trans. Nucl. Sci.* **2018**, *64*, 2511–2518. [CrossRef]
2. Raine, M.; Hubert, G.; Gaillardin, M.; Artola, L.; Paillet, P.; Girard, S.; Sauvestre, J.-E.; Bournel, A. Impact of the Radial Ionization Profile on SEE Prediction for SOI Transistors and SRAMs Beyond the 32-nm Technological Node. *IEEE Trans. Nucl. Sci.* **2011**, *58*, 840–847. [CrossRef]
3. Huang, P.; Chen, S.; Chen, J.; Liang, B.; Chi, Y. Heavy-Ion-Induced Charge Sharing Measurement with a Novel Uniform Vertical Inverter Chains (UniVIC) SEMT Test Structure. *IEEE Trans. Nucl. Sci.* **2015**, *62*, 3330–3338. [CrossRef]
4. Asenek, V.; Underwood, C.; Velazco, R. SEU induced errors observed in microprocessor systems. *IEEE Trans. Nucl. Sci.* **1998**, *45*, 2876–2883. [CrossRef]
5. Touloupis, E.; Flint, A.; Chouliaras, V.A. Study of the Effects of SEU-Induced Faults on a Pipeline-Protected Microprocessor. *IEEE Trans. Comput.* **2007**, *56*, 1585–1596.
6. Gao, J.; Li, Q. An SEU rate prediction method for microprocessors of space applications. *J. Nucl. Tech.* **2012**, *35*, 201–205.
7. Yu, C.Q.; Fan, L. A Prediction Technique of Single Event Effects on Complex Integrated Circuits. *J. Semicond.* **2015**, *36*, 115003-1–115003-5.
8. Buchner, S.P.; Miller, F.; Pouget, V.; McMorrow, D.P. Pulsed-Laser Testing for Single-Event Effects Investigations. *IEEE Trans. Nucl. Sci.* **2013**, *60*, 1852–1875. [CrossRef]
9. Shangguan, S.P. Experimental Study on Single Event Effect Pulsed Laser Simulation of New Material Devices. Ph.D. Thesis, National Space Science Center, The Chinese Academy of Sciences, Beijing, China, 2020.
10. Murphy, K.P. *Machine Learning: A Probabilistic Perspective*; MIT Press: Cambridge, MA, USA, 2012.
11. Hastie, T.; Tibshirani, R.; Friedman, J. *The Elements of Statistical Learning*, 2nd ed.; Springer: Berlin/Heidelberg, Germany, 2009.
12. Rousseeuw, P.J.; Hampel, F.R.; Ronchetti, E.M.; Stahel, W.A. *Robust Statistics: The Approach Based on Influence Functions*; John Wiley & Sons: Hoboken, NJ, USA, 2011.
13. Jun, S. *Mathematical Statistics*, 2nd ed.; Springer: Berlin/Heidelberg, Germany, 2003.

MDPI

St. Alban-Anlage 66

4052 Basel

Switzerland

Tel. +41 61 683 77 34

Fax +41 61 302 89 18

www.mdpi.com

Electronics Editorial Office

E-mail: electronics@mdpi.com

www.mdpi.com/journal/electronics